JERRY B. MARION

Professor of Physics
University of Maryland

RONALD C. DAVIDSON

Associate Professor of Physics
University of Maryland

SAUNDERS GOLDEN SERIES

MATHEMATICAL PREPARATION FOR GENERAL PHYSICS

W. B. SAUNDERS COMPANY

Philadelphia, London, Toronto

W. B. Saunders Company: West Washington Square
Philadelphia, Pa.   19105

1 St. Anne's Road
Eastbourne, East Sussex BN21 3UN, England

833 Oxford Street
Toronto, M8Z 5T9, Canada

Mathematical Preparation for General Physics   ISBN 0-7216-6070-3

Print No.:   9   8   7   6

INTRODUCTION

Students who enter a course in general physics not having had a recent course requiring mathematical skills frequently find themselves unable to keep pace because of a lack of familiarity with the necessary mathematical tools. Such students may experience considerable difficulty in rapidly acquiring the requisite mathematical background because of a lack of appropriate refresher courses in most colleges and universities or because no suitable (and brief) review book can be found. This book has been designed to fill the need of these students. We have assembled here a short overview of all the various types of mathematical topics that can reasonably be expected to be necessary in an introductory-level course in general physics. Because our intention is to assist the student in an *introductory* physics course, we have confined our discussions to *non*-calculus topics; the most advanced material in this book is concerned with the basic ideas of trigonometry and vector algebra.

We have attempted to present mathematical ideas from the standpoint of the *physicist*. That is, we are concerned here with operational mathematical techniques and not with mathematical rigor. Wherever possible we have introduced mathematical ideas with physical examples, and worked problems have frequently been included to emphasize each topic. In addition there are numerous exercises which the student should use to test his understanding of each new idea and technique. Answers to these exercises will be found in the back of this book, arranged in random-number fashion (to remove the annoyance of accidentally seeing the answer to the next exercise when checking the last result).

Because most physics courses begin with *mechanics,* all the examples used in the first eight chapters have been taken from this subject area. Additional examples from *electromagnetism* are used to supplement the material in the chapter that includes a discussion of electrical units.

While using this book the student should keep in mind that this is *not* a physics text—it is a mathematical aid and should be used in addition to a serious study of the assigned physics text.

JERRY B. MARION
RONALD C. DAVIDSON

College Park, Maryland

CONTENTS

8

9

APPENDIX A

APPENDIX B

DEALING WITH NUMBERS AND UNITS

The science of physics is based on performing and interpreting *experiments,* and the results of experiments can always be expressed in *numbers.* To deal with any physical situation in a meaningful way, we must be prepared to use and manipulate numbers, following the rules of mathematics. Indeed, mathematics is the natural language of physics and some appreciation of and familiarity with the techniques of elementary mathematics is necessary to make any real progress in the understanding of physics.

1.1 POWERS OF TEN

On the basis of certain measurements, we know that the distance from the Sun to the star Alpha Centauri is 4,070,000,000,000,000,000 centimeters. And we know that the mass of a single atom of hydrogen is 0.000 000 000 000 000 000 000 001 67 grams. This method of specifying the distance or the mass with a large number of zeroes is awkward and cumbersome. To overcome this difficulty in the expression of large or small numbers, we use a compact notation employing "powers of ten."

Multiplying 10 by itself a number of times, we find

$$10 \times 10 = 100 \qquad\qquad = 10^2$$

$$10 \times 10 \times 10 = 1000 \qquad\qquad = 10^3$$

$$10 \times 10 \times 10 \times 10 \times 10 = 100,000 = 10^5$$

The number of times that 10 is multiplied together (that is, the number of *zeroes* that appear in the answer) is used in the result as a superscript

1

to the 10. This superscript is called the *exponent* of 10 or the *power* to which 10 is raised. Also,

$$10^1 = 10$$

and, by convention,

$$10^0 = 1$$

If we express two numbers as powers of ten and then multiply these numbers, we have

$$10^2 \times 10^3 = (10 \times 10) \times (10 \times 10 \times 10) = 10^5 = 10^{(2+3)}$$

The general rule for this kind of operation is

$$10^n \times 10^m = 10^{(n+m)} \tag{1.1}$$

Just as 10^3 means $10 \times 10 \times 10$, we can raise *any* number to a power n by multiplying that number by itself n times. Thus,

$$2^3 = 2 \times 2 \times 2 = 8$$

$$4^2 = 4 \times 4 = 16$$

$$(1.5)^3 = 1.5 \times 1.5 \times 1.5 = 3.375$$

Similarly,

$$(10^2)^3 = (10^2) \times (10^2) \times (10^2)$$

$$= (10 \times 10) \times (10 \times 10) \times (10 \times 10)$$

$$= 10^6$$

The general rule for this kind of operation is

$$(10^n)^m = 10^{(n \times m)} \tag{1.2}$$

EXERCISES

1. $10^9 = $ ____   (Ans.   87)

2. $10{,}000{,}000 = $ ____   (Ans.   75)

3. $(10^3) \times (10^5) = $ ____   (Ans.   102)

4. $3^4 = $ ____   (Ans.   21)

5. $(1.2)^3 = $ ____   (Ans.   98)

6. $(1000)^4 = $ ____   (Ans.   115)

1.2   NEGATIVE EXPONENTS

If a power of 10 appears in the denominator of an expression, the exponent is given a negative sign:

$$\frac{1}{10} = 0.1 = 10^{-1}$$

$$\frac{1}{1000} = 0.001 = \frac{1}{10^3} = 10^{-3}$$

In general,

$$\frac{1}{10^m} = 10^{-m} \qquad (1.3)$$

Multiplying a *positive* power of 10 by a *negative* power of 10 gives

$$10{,}000 \times 0.01 = 10^4 \times \frac{1}{10^2} = 10^4 \times 10^{-2} = 10^{(4-2)} = 10^2$$

$$\frac{100}{1000} = \frac{10^2}{10^3} = 10^{(2-3)} = 10^{-1} = 0.1$$

In general, combining the rules given in Equations 1.1 and 1.3, we have

$$\frac{10^n}{10^m} = 10^n \times 10^{-m} = 10^{(n-m)} \qquad (1.4)$$

The same rules for the use of negative exponents apply for numbers other than 10. For example,

$$2^{-2} = \frac{1}{2^2} = \frac{1}{4} = 0.25$$

$$\frac{3^4}{3^2} = 3^{(4-2)} = 3^2 = 9$$

$$\frac{2^3 \times 3^5}{2^2 \times 3^2} = 2^{(3-2)} \times 3^{(5-2)} = 2^1 \times 3^3 = 2 \times 27 = 54$$

EXERCISES

1. $\dfrac{1}{100{,}000} = $ ____   (Ans. 293)

2. $0.000\ 001 = $ ____   (Ans. 104)

3. $\dfrac{10^6}{1000} = $ ____   (Ans. 282)

4. $\dfrac{0.001}{10^4} =$ ___ (Ans. 195)

5. $\dfrac{10^3 \times 0.1}{10,000 \times 10^{-2}} =$ ___ (Ans. 93)

6. $\dfrac{4^3 \times 3^4}{2^4 \times 3^3} =$ ___ (Ans. 322)

1.3 PREFIXES

When discussing physical quantities, it frequently proves convenient to use a *prefix* to a unit instead of a power of 10. For example, *centi-* means $\dfrac{1}{100}$, so *centi*-meter or centimeter (cm) means $\dfrac{1}{100}$ of a meter (m):

$$1 \text{ m} = 100 \text{ cm} = 10^2 \text{ cm}; \quad 1 \text{ cm} = 0.01 \text{ m} = 10^{-2} \text{ m}$$

Similarly, *milli-* means $\dfrac{1}{1000} = 10^{-3}$ and *mega-* means 10^6:

$$1 \text{ m} = 10^3 \text{ millimeters (mm)}; \quad 1 \text{ mm} = 10^{-3} \text{ m}$$

$$\$1,000,000 = 1 \text{ megabuck}$$

Table 1.1 lists some of the most frequently used prefixes.

TABLE 1.1 PREFIXES EQUIVALENT TO POWERS OF 10

Prefix	Symbol	Power of 10
giga-	G	10^9 *
mega-	M	10^6 *
kilo-	k	10^3
centi-	c	10^{-2}
milli-	m	10^{-3}
micro-	μ	10^{-6}
nano-	n	10^{-9}
pico-	p	10^{-12}
femto-	f	10^{-15}

*$10^6 = 1$ *million*. In the U.S., $10^9 = 1$ *billion*, but the European convention is that $10^9 = 1000$ million and that 1 billion $= 10^{12}$; the prefix *giga-* is internationally agreed on to represent 10^9.

EXERCISES

1. 1 kilometer (km) = ___ m (Ans. 146)

2. 1 nanosecond (ns) = ___ s (Ans. 182)

3. 1 mm = ___ km (Ans. 91)

4. (1 km) × (1 m) = ___ cm² (Ans. 310)

1.4 CALCULATIONS WITH POWERS OF TEN

By using powers of 10, many types of calculations are made considerably easier. First, we note that *any* number can be expressed in terms of a power of 10 by writing, for example,

$$6400 = 6.4 \times 10^3$$

$$0.0137 = 1.37 \times 10^{-2}$$

$$970,000 = 0.97 \times 10^6$$

Notice that in changing from an ordinary number to a number expressed as a power of 10, the exponent of 10 corresponds to the number of places that the decimal has been moved. (A positive exponent means that the decimal has been moved to the left and a negative exponent means that the decimal has been moved to the right.) Thus, in expressing 6400 in terms of a power of 10, we move the decimal three places to the left to obtain 6.4×10^3. And in expressing 0.0137 in terms of a power of 10, we move the decimal two places to the right, obtaining 1.37×10^{-2}.

Usually, we express a quantity in this notation by writing the multiplying factor as a number between 0.1 and 10. That is, we write $240,000 = 2.4 \times 10^5$ instead of 240×10^3. (See, however, the discussion in Section 2.3.)

$$\frac{42,000,000}{3000} = \frac{4.2 \times 10^7}{3 \times 10^3} = \frac{4.2}{3} \times \frac{10^7}{10^3} = 1.4 \times 10^4$$

$$0.0012 \times 0.000\,003 = (1.2 \times 10^{-3}) \times (3 \times 10^{-6})$$

$$= (1.2 \times 3) \times (10^{-3} \times 10^{-6}) = 3.6 \times 10^{-9}$$

EXERCISES

1. Express the distance from the Sun to Alpha Centauri (in cm) as a power of 10. (See Section 1.1.)   (Ans. 183)

2. Express the mass of a hydrogen atom (in grams) as a power of 10. (See Section 1.1.)   (Ans. 250)

3. $\dfrac{6,400,000}{1600} = $ ____   (Ans. 232)

4. $3200 \times 0.0004 = $ ____   (Ans. 355)

5. $\dfrac{24,000}{0.012} = $ ____   (Ans. 101)

6. $\dfrac{160 \times 0.0024}{0.32 \times 480,000} = $ ____   (Ans. 323)

7. $16,000 \times 0.03 \times 0.12 = $ ____   (Ans. 110)

1.5 THINKING IN ORDERS OF MAGNITUDE

One power of 10 is sometimes referred to as an *order of magnitude*. Thus, "a dollar is an order of magnitude more valuable than a dime and two orders of magnitude (a factor of 100) more valuable than a penny." More frequently, the term *order of magnitude* is applied in an approximate sense. Thus, "the Earth is two orders of magnitude more massive than the Moon." (Actually, the mass of the Earth is 81 times the mass of the Moon.) Or, since the meter is 39.37 inches and the yard is 36 inches, we say "the meter and the yard are of the same order of magnitude."

Because we deal so frequently with physical quantities that are very large or very small, it is extremely helpful if we cultivate the habit of thinking about and referring to these quantities in orders of magnitude. Even if we do not know the precise value of a certain quantity it is still useful to know the order of magnitude of this size. For example, the diameter of an atom is of the order of 10^{-8} cm. If we perform some calculation and obtain a result which says that the size of an atom is 10^{-5} cm, then we know immediately that something is wrong in the calculation!

We will use the symbol $\sim$ to indicate "is of the order of magnitude of." Often we include one number in addition to the power of 10, so that a somewhat more precise (but still approximate) value results. Some typical order-of-magnitude quantities are as follows:

(a) There are $\sim 3 \times 10^7$ seconds in a year.

(b) Light travels ~ 1 foot in 10^{-9} seconds.

(c) The diameter of a nucleus is $\sim 10^{-4}$ of the diameter of an atom.

(d) The mass of the Sun is $\sim 3 \times 10^5$ times the mass of the Earth.

(e) The mass of a hydrogen atom is $\sim 2 \times 10^3$ times the mass of an electron.

None of the values given above is *precise*, but an appreciation of the *approximate* value of a quantity permits the rapid evaluation of a situation and can often be used to determine whether a particular calculation is reliable.

EXERCISES

Give order-of-magnitude answers for the following:

1. The Universe is thought to be $\sim 3 \times 10^{10}$ years old. Express this age in seconds.   (Ans. 351)

2. The mass of the Sun is $\sim 2 \times 10^{33}$ grams and consists mainly of hydrogen. Approximately how many atoms of hydrogen are there in the Sun? (Refer to the mass of the hydrogen atom given in Section 1.1)   (Ans. 227)

3. The area of Maryland is 10,577 square miles

and the population is $\sim 3.8 \times 10^6$. What is the population density (persons per square mile) in Maryland?

(Ans. 199)

4. Express your age in seconds.

1.6 FRACTIONAL EXPONENTS

An *integer* exponent represents the *power* to which a number is raised; for example, $10^3 = 10 \times 10 \times 10 = 1000$ or $3^2 = 3 \times 3 = 9$. *Fractional* exponents also have meaning and these exponents are manipulated according to the same rules as for integer exponents. For example,

$$4^{\frac{1}{2}} \times 4^{\frac{1}{2}} = 4^{\left(\frac{1}{2}+\frac{1}{2}\right)} = 4^1 = 4, \text{ or } (4^{\frac{1}{2}})^2 = 4$$

$$27^{\frac{1}{3}} \times 27^{\frac{1}{3}} \times 27^{\frac{1}{3}} = 27^{\left(\frac{1}{3}+\frac{1}{3}+\frac{1}{3}\right)} = 27^1 = 27, \text{ or } (27^{\frac{1}{3}})^3 = 27$$

What is the meaning of a fractional exponent? In the examples above, notice that if a number $N^{1/m}$ is raised to the power m, the result is N itself. We call $N^{1/m}$ the mth *root* of N, and we also use the equivalent notation,

$$\sqrt[m]{N} = N^{1/m} \tag{1.5}$$

If $m = 2$, this factor is usually suppressed from the radical sign, $\sqrt{}$. For example,

$$4 \times 4 = 4^2 = 16, \text{ so that } \sqrt{16} = 16^{\frac{1}{2}} = 4$$

$$\sqrt{5^2} = (5^2)^{\frac{1}{2}} = 5^{\left(2 \times \frac{1}{2}\right)} = 5^1 = 5$$

$$2 \times 2 \times 2 = 2^3 = 8, \text{ so that } \sqrt[3]{8} = 8^{\frac{1}{3}} = 2$$

This latter result can also be obtained by writing

$$\sqrt[3]{8} = \sqrt[3]{2^3} = (2^3)^{\frac{1}{3}} = 2^{\left(3 \times \frac{1}{3}\right)} = 2^1 = 2$$

More complicated fractional exponents are handled in the following way:

$$4^{\frac{3}{2}} = 4^{\left(\frac{1}{2} \times 3\right)} = (4^{\frac{1}{2}})^3 = (2)^3 = 8$$

or,

$$4^{\frac{3}{2}} = 4^{\left(3 \times \frac{1}{2}\right)} = (4^3)^{\frac{1}{2}} = (64)^{\frac{1}{2}} = \sqrt{64} = 8$$

Notice that the *numerator* of the fractional exponent always represents a *power* to which the number is raised and the *denominator* always represents the *root* of the number.

$$\sqrt[3]{8000} = (8000)^{\frac{1}{3}} = (2^3 \times 10^3)^{\frac{1}{3}} = 2 \times 10 = 20$$

$$(2.7 \times 10^4)^{\frac{1}{3}} = (27,000)^{\frac{1}{3}} = (27 \times 10^3)^{\frac{1}{3}} = \sqrt[3]{27} \times 10 = 3 \times 10 = 30$$

$$(64,000)^{\frac{2}{3}} = (64 \times 10^3)^{\frac{2}{3}} = (\sqrt[3]{64})^2 \times 10^{(3 \times \frac{2}{3})}$$

$$= 4^2 \times 10^2 = 1600$$

In writing equations such as $\sqrt{4} = 2$, we are not being properly general. The equation $\sqrt{4} = 2$ means that $2^2 = 2 \times 2 = 4$. But it is also true that $(-2) \times (-2) = 4$, so we could equally well write $\sqrt{4} = -2$. There is always an ambiguity of sign in extracting roots. We will usually want the *positive* sign (the *physics* of the problem will dictate which sign is physically meaningful), but we should properly indicate the possibility of either sign by writing $\sqrt{4} = \pm 2$. (The symbol $\pm$ means "positive or negative.")

EXERCISES

1. $27^{\frac{1}{3}} = $ ____ (Ans. 109)

2. $64^{\frac{1}{2}} = $ ____ (Ans. 155)

3. $16^{\frac{1}{4}} = $ ____ (Ans. 262)

4. $16^{\frac{3}{2}} = $ ____ (Ans. 94)

5. $\left(\dfrac{32 \times 10^5}{2^3 \times 10^3}\right)^{\frac{1}{2}} = $ ____ (Ans. 97)

6. $\left[10 \times \left(\dfrac{10.8 \times 10^3}{300}\right)^{\frac{1}{2}} + 4\right]^{\frac{1}{3}} = $ ____ (Ans. 20)

1.7 MATHEMATICAL NOTATION

In ordinary equations we use the symbol $=$ to denote equality of two quantities:

$$y = 16.27\, t^2 \quad \text{or} \quad A \times B = C$$

Even if we did not know the factor 16.27 which occurs in the above equation, we could still state that y is *proportional to* t^2, and we would write

$$y = kt^2 \quad \text{or} \quad y \propto t^2$$

Or, if we knew only that the factor is *approximately* equal to 16, we would write

$$y \cong 16\, t^2$$

We have already seen that the symbol $\sim$ (or $\approx$) is used to indicate that two quantities are *very approximately* equal or have the same *order of magnitude*.

The symbols $<$ and $>$ mean, respectively, *is less than* and *is greater than*; for example,

$$\text{area of Canada} > \text{area of Argentina}$$

$$\text{mass of the Earth} < \text{mass of Jupiter}$$

If a quantity is *very much smaller* or *very much larger* than another quantity, we use a double symbol:

$$\text{area of Canada} \gg \text{area of Luxembourg}$$

$$\text{mass of the Earth} \ll \text{mass of the Milky Way Galaxy}$$

Sometimes we know only that a quantity is smaller or larger than some poorly defined quantity. In such a case we use the symbol $\lesssim$ (or $\gtrsim$) to denote *is less than about* (or *is greater than about*). For example,

$$\text{population of the U.S.} \gtrsim 2 \times 10^8$$

We frequently find it convenient to use a shorthand notation to indicate the *change* in the value of a quantity. If an object is located at the position $x_1 = 2$ cm at a certain time and if at a later time the location is $x_2 = 9$ cm, we say that the distance moved (or the change in x) is $x_2 - x_1 = 9$ cm $- 2$ cm $= 7$ cm. That is, we take the *final* position (x_2) and subtract from it the *initial* position (x_1). This change in x is often denoted by the symbol Δx:

$$\Delta x = x_{\text{final}} - x_{\text{initial}} = x_2 - x_1 \tag{1.6}$$

The symbol Δx does *not* imply the product of Δ and x, but means "the change in x" or "an increment of x." In general, a Greek delta, Δ, in front of a quantity means the *change* in that quantity; e.g., $t_2 - t_1 = \Delta t =$ time difference. Δt can be either *positive* ($\Delta t > 0$) or *negative* ($\Delta t < 0$), depending on whether t_2 is greater or smaller than t_1.

Sometimes we are interested only in the *magnitude* of a quantity and not whether it carries a positive or a negative sign. We denote the magnitude of a quantity by vertical bars, $|x|$. That is, for positive values of x, $|+x| = x$ and $|-x| = x$.

$$|3x - 7x| = |-4x| = 4x$$

The meanings of the symbols we will find useful in this book are summarized in Table 1.2.

TABLE 1.2   MATHEMATICAL SYMBOLS AND THEIR MEANINGS

Symbol	Meaning
$=$	is equal to
$\propto$	is proportional to
$\cong$	is approximately equal to
$\approx$ or $\sim$	is *very* approximately equal to; is of the order of magnitude of
$>$ ($<$)	is greater (less) than
$\gg$ ($\ll$)	is much greater (less) than
$\gtrsim$ ($\lesssim$)	is greater (less) than about
Δx	change in x
$\lvert x \rvert$	magnitude of x

EXERCISES

Insert the appropriate symbol between the following pairs of quantities (in some cases there may be two appropriate symbols — give both):

1. height of Mont Blanc   Mt. Everest   (Ans. 143)

2. area of Canada   area of Brazil   (Ans. 118)

3. Gross National Product 10^3 gigabucks   (Ans. 357)

4. mass of an apple   mass of an orange   (Ans. 308)

5. If $\dfrac{a}{b} = 6$, then a b (Ans. 327)

1.8 FUNDAMENTAL PHYSICAL QUANTITIES

Physical quantities not only have *magnitudes*, specified by *numbers*, but also *dimensions* or *units*. It makes no sense to say that a certain length is "75" unless we also state the appropriate units — feet, meters, miles, or whatever. Although we encounter a wide variety of physical quantities that require units for their complete specification — for example, distance, force, energy, momentum, and electric field strength — these various units can be expressed in terms of only *three* fundamental quantities. The basic units of physical measure are those of *length*, *time*, and *mass* — the dimensions of all other physical quantities can be expressed in terms of the units of these three. For example, *speed* is measured by the distance traveled in a certain time and so the dimensions of speed are *length/time*. In the next section and in Chapter 9 we will review the dimensions of other physical quantities.

The system of units most widely used in physics is the *metric system*. Within the metric system there are two variations: the MKS system in which the standard units are the meter, the kilogram, and the second; and the CGS system in which the standard units are the centimeter, the gram, and the second. (In terms of the fundamental units of length, mass, and time, the MKS and CGS units differ only by simple factors of 10, but when electrical units are considered, the relationship between the two systems becomes more complicated; see Chapter 9.) Also in use at the present time is the *British engineering system* in which the standard units are the yard, the pound-mass, and the second. It seems inevitable that the metric system will replace the British system and will become the international standard system within a matter of years.

The fundamental units of measure in the metric and British systems are summarized in Table 1.3.

If we wish to convert a certain length from one system of units to another, we use the following procedure. Since 1 in = 2.54 cm, we can form the ratios

$$\frac{2.54 \text{ cm}}{1 \text{ in}} = 1, \quad \text{or} \quad \frac{1 \text{ in}}{2.54 \text{ cm}} = 1$$

These ratios are just *unity*, so we can multiply any quantity by either ratio without changing the value. Thus,

$$15 \text{ in} = (15 \text{ in}) \times \left(\frac{2.54 \text{ cm}}{1 \text{ in}}\right) = 38.10 \frac{\text{cm-in}}{\text{in}} = 38.10 \text{ cm}$$

where the *inches* in numerator and denominator cancel. Also,

$$48 \text{ cm} = (48 \text{ cm}) \times \left(\frac{1 \text{ in}}{2.54 \text{ cm}}\right) = 18.9 \text{ in}$$

$$36 \text{ kg-m} = (36 \text{ kg-m}) \times \left(\frac{10^3 \text{ g}}{1 \text{ kg}}\right) \times \left(\frac{10^2 \text{ cm}}{1 \text{ m}}\right) = 3.6 \times 10^6 \text{ g-cm}$$

$$60 \text{ mi/hr} = (60 \text{ mi/hr}) \times \left(\frac{5280 \text{ ft}}{1 \text{ mi}}\right) \times \left(\frac{1 \text{ hr}}{3600 \text{ s}}\right) = 88 \text{ ft/s}$$

We will rarely use British units except in a few elementary examples involving the motion of objects.

TABLE 1.3 FUNDAMENTAL UNITS OF MEASURE

	Metric		British
	MKS	*CGS*	
Length	Meter (m) 1 m = 100 cm	Centimeter (cm)	Yard (yd) 1 yd = 36 in 1 in = 2.54 cm
Mass	Kilogram (kg) 1 kg = 1000 g	Gram (g)	Pound-mass (lb) 1 lb = 453.59 g
Time	Second (s)	Second (s)	Second (s)

Any quantity which does not have units is said to be *dimensionless* or a *pure number*. For example, the *ratio* of two physical quantities with the same dimensions is *dimensionless*:

$$R = \frac{8 \text{ cm}}{4 \text{ cm}} = 2$$

The quantity R is a *pure number*.

EXERCISES

1. 1 km = _____ cm   (Ans. 274)

2. 3 lb = _____ kg   (Ans. 242)

3. 43 mm = _____ in   (Ans. 121)

4. 1 km = _____ mi   (Ans. 312)

5. 7.2 yd/s = _____ m/s   (Ans. 277)

6. 14 m/s = _____ km/hr   (Ans. 162)

7. 107 lb-in = _____ g-cm   (Ans.   33)

8. 1 year = _____ s   (Ans. 244)

1.9 THE DIMENSIONS OF PHYSICAL QUANTITIES

Although length, mass, and time are the *fundamental* physical quantities, there is a variety of additional concepts that are useful and important in physics. In later chapters we will refer to some of these concepts when presenting physical examples of mathematical ideas. Therefore, we summarize here the important physical quantities that appear in the area of *mechanics*. Even though the dimensions of each of these quantities can be expressed in terms of length, mass, and time, it is customary to attach special names to the units for certain quantities. In such cases, the dimensions are given in terms of the fundamental units and in terms of the *derived* unit.

Velocity (v). Speed, or the rate at which position changes with time:*

$$v = \frac{\Delta x}{\Delta t}, \text{ m/s, cm/s, mi/hr, etc.} \tag{1.7}$$

Acceleration (a). Speeding up or slowing down; the rate at which velocity changes with time:

---

*The terms *velocity* and *speed* are similar in meaning, but as used in physics there is an important distinction. The concept of motion involves not only the *rate* at which an object moves but also the *direction* of movement. The term *velocity* is usually reserved for use when the direction of the motion is important (that is, when the *vector* character of velocity is important; see Chapter 7). The term *speed* is used when only the rate of movement (regardless of direction) is to be specified.

$$a = \frac{\Delta v}{\Delta t}, \text{ (m/s)/s or m/s}^2, \text{ cm/s}^2, \text{ ft/s}^2, \text{ etc.} \tag{1.8}$$

Momentum (p). The product of a body's mass and its velocity:

$$p = mv, \text{ kg-m/s, g-cm/s} \tag{1.9}$$

Force (F). A push or a pull; the fundamental definition is in terms of the rate at which a body's momentum changes with time:

$$F = \frac{\Delta p}{\Delta t}, \text{ kg-m/s}^2, \text{ g-cm/s}^2 \tag{1.10}$$

Another important equation for force is Newton's second law:

$$F = ma, \text{ kg-m/s}^2, \text{ g-cm/s}^2 \tag{1.11}$$

The units of force in the MKS and CGS systems are

$$\text{MKS: 1 kg-m/s}^2 = 1 \text{ newton (N)} \tag{1.12}$$

$$\text{CGS: 1 g-cm/s}^2 = 1 \text{ dyne} \tag{1.13}$$

Since 1 kg $= 10^3$ g and 1 m $= 10^2$ cm, the relationship between dynes and newtons is

$$1 \text{ N} = 10^5 \text{ dynes} \tag{1.14}$$

Work (W). The expenditure of energy; the product of force (F) and the distance (s) through which the force acts:

$$W = Fs, \text{ kg-m}^2/\text{s}^2, \text{ g-cm}^2/\text{s}^2 \tag{1.15}$$

The units of work in the MKS and CGS systems are

$$\text{MKS: 1 kg-m}^2/\text{s}^2 = 1 \text{ N-m} = 1 \text{ joule (J)} \tag{1.16}$$

$$\text{CGS: 1 g-cm}^2/\text{s}^2 = 1 \text{ dyne-cm} = 1 \text{ erg} \tag{1.17}$$

The relationship between ergs and joules is (verify this)

$$1 \text{ J} = 10^7 \text{ ergs} \tag{1.18}$$

Energy (E). The capacity to do work; energy comes in many forms—in mechanical situations we identify *kinetic* energy (the energy possessed by a body because of its *motion*) and *potential* energy (the energy possessed by a body because of its *position*):

Kinetic energy: $E_K = \frac{1}{2} mv^2$, kg-m²/s², g-cm²/s², $\qquad$ (1.19)

or J, ergs

For the particular case of a body of mass m in the gravitational field near the surface of the Earth, the potential energy is

Potential energy: $E_P = mgh$, kg-m²/s², g-cm²/s², $\qquad$ (1.20)

or J, ergs

where g is the acceleration due to gravity (i.e., the acceleration experienced by a freely falling body) and where h is the height of the body above some reference position (e.g., the surface of the Earth.) Near the surface of the Earth, g has the value

$$g \cong 9.8 \text{ m/s}^2$$
$$\cong 980 \text{ cm/s}^2 \qquad (1.21)$$
$$\cong 32 \text{ ft/s}^2$$

Notice that *work* and *energy* have the *same* dimensions. Table 1.4 summarizes the units of these various mechanical quantities.

TABLE 1.4 SOME PHYSICAL QUANTITIES AND THEIR UNITS

Quantity	Equation(s)	Units	
		MKS	*CGS*
Velocity (or speed°)	$v = \dfrac{\Delta x}{\Delta t}$	m/s	cm/s
Acceleration	$a = \dfrac{\Delta v}{\Delta t}$	m/s²	cm/s²
Momentum	$p = mv$	kg-m/s	g-cm/s
Force	$F = \dfrac{\Delta p}{\Delta t}$	kg-m/s²	g-cm/s²
	$F = ma$	= newton(N)	= dyne
Work,	$W = Fs$	kg-m²/s²	g-cm²/s²
Energy	$E_K = \frac{1}{2} mv^2$	= N-m	= dyne-cm
	$E_P = mgh$	= joule(J)	= erg

1 N = 10⁵ dynes
1 J = 10⁷ ergs

°See footnote on p. 12.

Example 1.9.1

A particle has a velocity of 20 ft/s at $t_1 = 4.5$ s and a velocity of 48 ft/s at $t_2 = 6.8$ s. Express the acceleration in m/s²:

First, we convert the velocities to m/s:

$$v_1 = 20 \text{ ft/s} = \left(20 \frac{\text{ft}}{\text{s}}\right) \times \left(\frac{12 \text{ in}}{1 \text{ ft}}\right) \times \left(\frac{2.54 \text{ cm}}{1 \text{ in}}\right) \times \left(\frac{1 \text{ m}}{100 \text{ cm}}\right)$$

$$= 6.1 \text{ m/s}$$

$$v_2 = 48 \text{ ft/s} = \left(48 \frac{\text{ft}}{\text{s}}\right) \times \left(\frac{12 \text{ in}}{1 \text{ ft}}\right) \times \left(\frac{2.54 \text{ cm}}{1 \text{ in}}\right) \times \left(\frac{1 \text{ m}}{100 \text{ cm}}\right)$$

$$= 14.6 \text{ m/s}$$

Then,

$$a = \frac{\Delta v}{\Delta t} = \frac{v_2 - v_1}{t_2 - t_1} = \frac{14.6 \text{ m/s} - 6.1 \text{ m/s}}{6.8 \text{ s} - 4.5 \text{ s}}$$

$$= \frac{8.5 \text{ m/s}}{2.3 \text{ s}} = 3.7 \text{ m/s}^2$$

The sign of a is *positive*, so the particle was *speeding up*; if v_1 had exceeded v_2, the sign of a would have been *negative*, indicating that the particle was *slowing down*.

Example 1.9.2

What is the kinetic energy (in joules) of a 800-g particle that is moving with a velocity of 900 cm/s?

First, we convert the mass and the velocity to MKS units:

$$m = 800 \text{ g} = (800 \text{ g}) \times \left(\frac{1 \text{ kg}}{1000 \text{ g}}\right) = 0.8 \text{ kg}$$

$$v = 900 \text{ cm/s} = (900 \text{ cm/s}) \times \left(\frac{1 \text{ m}}{100 \text{ cm}}\right) = 9 \text{ m/s}$$

Then,

$$E_k = \frac{1}{2} m v^2 = \frac{1}{2} \times (0.8 \text{ kg}) \times (9 \text{ m/s})^2$$

$$= 36 \text{ J}$$

EXERCISES

1. Newton's law of universal gravitation states that the gravitational force between two bodies with masses m_1 and m_2 which are separated by a distance r is $F = Gm_1m_2/r^2$, where G is a constant. What are the dimensions of G in MKS and in CGS units? (Ans.   57)

2. A constant force of $5\ N$ acts on a body and the body moves a distance of 3 m. Express the work done in *ergs*. (Ans.   141)

3. A 400-g object is moving with a velocity of 200 cm/s. What is the kinetic energy of the object in *joules*? (Ans.   35)

4. What force (in newtons) is required to change the momentum of a body from 20 kg-m/s to 80 kg-m/s in a period of 3 s? (Ans.   22)

5. A 2-kg object is moved from the surface of the Earth to a height of 40 m. What is the change in potential energy of the object? (Ans.   56)

CHAPTER 2

EXPERIMENTAL ERROR AND SIGNIFICANT FIGURES

2.1 EXPERIMENTAL NUMBERS— PROBABLE ERROR

In physics we seek to provide an orderly and precise description of natural phenomena through the use of *numbers*. In the report of an experiment, the use of numbers instead of qualitative descriptions is always preferred. For example, suppose that we examine the light emitted in a certain physical process. We could report the result of the experiment by saying "The color of the light is yellow." But we give a much more precise statement of the result if we say "The emitted light has a wavelength of 5.236×10^{-7} m."

More is involved in reporting the result of an experiment than a simple statement of the experimental number. To state that "$\lambda = 5.236 \times 10^{-7}$ m" leaves unanswered the question, "How precise is the result?" Could the wavelength actually be 5.300×10^{-7} m? Could the calibration of the measuring instrument be in error by an amount sufficient to yield a wavelength of 5.236×10^{-7} m when the true wavelength is 5.300×10^{-7} m? An experimenter must always carefully analyze his result in terms of the measuring standards used, the reliability of his equipment, and any factors which could influence the result. After he has done this and after he has measured the quantity many times to check the reproducibility of his instruments and his technique, the experimenter will be in a position to state the result in the following form:

$$\lambda = (5.236 \pm 0.002) \times 10^{-7} \text{ m}$$

An experimental result presented in this way means that the experimenter has measured the wavelength many times and has found a *mean* or *average* value of 5.236×10^{-7} m. Furthermore, he has carefully examined the calibration and reliability of his equipment and has assigned a *probable error* of 0.002×10^{-7} m to his result. "Probable error" means that if additional measurements are made — either by the same experimenter with the same equipment or by a different experimenter with different equipment — the chance that the result will lie *within* the range of the probable error is the same as the chance that the result will lie *outside* the range of the probable error. That is, there is a 50 per cent probability that a new measurement will lie in the range from $(5.236 - 0.002) \times 10^{-7}$ m $= 5.234 \times 10^{-7}$ m to $(5.236 + 0.002) \times 10^{-7}$ m $= 5.238 \times 10^{-7}$ m. If the result of a new experiment is $(5.237 \pm 0.002) \times 10^{-7}$ m, we would say that the two results are in *agreement*, since each falls within the assigned probable error of the other. On the other hand, if the new result is $(5.241 \pm 0.001) \times 10^{-7}$ m, we would say that the two experiments *disagree*. If we were forced to choose one result over the other, we would choose the second value because it has a smaller probable error and therefore should represent a more reliable result.

Experimental results (and their corresponding probable errors) are often plotted in graphs. The method of displaying the results is illustrated in Figure 2.1 where the three values mentioned in the preceding paragraph are shown. The probable errors are indicated by the *error bars*, the horizontal lines extending in both directions from the central dot which represents the experimental (mean) value.

The assignment of a *probable error* to a result is the experimenter's way of stating the *precision* of the result. When we measure some physical quantity, we do not know (nor *can* we know) the *true* value of the quantity. Therefore, we can never know the *error* in an experimental result. We can only analyze the experiment and state the assessment of reliability — the *precision* — in terms of the *probable error*. (It is unfortunate that the term "probable error" is used in this way. This quantity does not represent the most probable deviation from the actual value; instead, it means that there are uncertainties in the measurement and that the true value probably lies within the range indicated.)

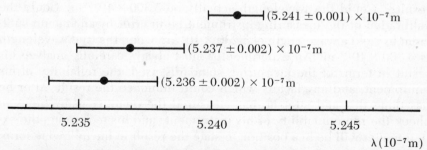

FIGURE 2.1 Three experimental results and their probable errors plotted on a wavelength scale.

EXERCISE

The following are "experimental" values of π, determined by several individuals who measured the circumference and the diameter of a circle and calculated the ratio: 3.141 ± 0.001, 3.144 ± 0.002, 3.140 ± 0.002, 3.15 ± 0.01, 3.1416 ± 0.0005. Which of the results is in agreement with the true value? (Notice that the measurements here are of a *mathematical* quantity, not a *physical* quantity, so a "true" value does exist.) If you were to choose one of the values as the most reliable, which would it be?

(Ans. 49)

2.2 ABSOLUTE AND RELATIVE ERROR

If the result of a certain length measurement is stated as $L = 3.08 \pm 0.03$ cm, we mean that the *absolute error* in L is ±0.03 cm. (Actually, we should use the term *absolute probable error*, but the "probable" is usually omitted in the interests of brevity.) The error is "absolute" in the sense that it is stated in the same physical units as the quantity measured. The *relative error* in L is the fraction of L represented by the absolute error. Let the absolute error in L be ΔL; then, the specification of a measurement of L is given by a statement of the form $L \pm \Delta L$. Thus,

$$\text{Relative error in } L = \frac{\Delta L}{L} \qquad (2.1)$$

For example, if $L \pm \Delta L = 3.08 \pm 0.03$ cm, then the absolute and relative errors are

$$\text{Absolute error} = \Delta L = \pm 0.03 \text{ cm}$$

$$\text{Relative error} = \frac{\Delta L}{L} = \pm \frac{0.03 \text{ cm}}{3.08 \text{ cm}} = \pm 0.01 \ (\pm 1\%)$$

That is, the relative error in L is 0.01 or 1% of the value of L. The smaller the relative error, the more precise is the measurement. But a smaller *absolute* error does not necessarily mean an improvement in the relative error. For example, the result $L' = 1.07 \pm 0.02$ cm has a smaller absolute error than that assigned to L above, but the relative error in L' is 2% and so this measurement is less precise in a relative sense.

Very precise measurements sometimes carry relative errors that are given in *parts per million* (ppm). A relative error of 1 ppm is $\pm0.000\,001$ ($\pm10^{-4}$ %).

Occasionally one sees a result stated as $(3.98 \times 10^4 \text{ cm}) \pm 2\%$. This means that the relative error is 0.02, and if the absolute error is desired, it must be calculated: in this case, absolute error $= 0.02 \times (3.98 \times 10^4$ cm$) = 0.08 \times 10^4$ cm $= 800$ cm.

Example 2.2.1

What is the precision of our present knowledge of the velocity of light?

According to a recent careful study of all the various measurements of the velocity of light, the value is

$$c = (2.997\ 925 \pm 0.000\ 001) \times 10^8 \text{ m/s}$$

Absolute error $= 0.000\ 001 \times 10^8$ m/s $= 100$ m/s

$$\text{Relative error} = \frac{0.000\ 001 \times 10^8 \text{ m/s}}{2.997\ 925 \times 10^8 \text{ m/s}} = 0.000\ 000\ 33$$

$$= 0.33 \text{ ppm}$$

The velocity of light is the most precisely known of all the fundamental physical constants.

EXERCISES

What is the relative error in each of the following values:

1. 3.97 ± 0.02 cm   (Ans. 208)

2. 4.868 ± 0.005 m/s   (Ans. 220)

3. $(3.715 \pm 0.003) \times 10^{11}$ m   (Ans. 213)

What is the relative error (in ppm) in each of the following values:

4. $7.848\ 632 \pm 0.000\ 008$ g   (Ans.   54)

5. $2.146\ 8756 \pm 0.000\ 0003$ (Ans. 131)

6. $(4.8675 \pm 0.0002) \times 10^{-8}$ cm   (Ans. 269)

2.3 SIGNIFICANT FIGURES

When we refer to an experimental result, we often do not give explicitly the probable error. That is, instead of writing $L = 3.264 \pm 0.002$ m, we write only $L = 3.264$ m. We imply by this shorthand notation that the uncertainty in the result is in the *last digit* given (but we do not know whether the uncertainty is ± 1 digit, ± 2 digits, or ± 5 digits). Thus, if we write $L = 3.264$ m, we imply that the 4 is uncertain but not the 6. If the result were $L = 3.264 \pm 0.012$ m, we would write the shortened value as $L = 3.26$ m, not as $L = 3.264$ m, because now the uncertainty is in the second decimal instead of the third. In this shorthand method we give only the *significant figures* of the result.

Notice that we always give more information by explicitly writing the probable error associated with a result than by merely using the significant figures. We can only *estimate* the probable error in a result from a statement of the significant figures. For example, each of the results with probable errors listed below has the same shorthand value in terms of the significant figures:

$$\left.\begin{array}{l} 2.682 \ \pm 0.002 \text{ m} \\ 2.682 \ \pm 0.004 \text{ m} \\ 2.6821 \pm 0.0016 \text{ m} \\ 2.6821 \pm 0.0009 \text{ m} \end{array}\right\} 2.682 \text{ m}$$

Notice in the last case that, although the probable error is in the fourth decimal, ±0.0009 m, we *round-off* the result to 2.682 m because the error is closer to ±0.001 m than to ±0.0001 m. In general, we round-off numbers *upwards* if the last digit is greater than 5, and we round-off downwards if the last digit is less than 5. If the last digit is equal to 5, by convention we round-off upwards if the next-to-last digit is *even* and we round-off downwards if the next-to-last digit is *odd*. For example,

$$7.687 \pm 0.012 \text{ m} \rightarrow 7.69 \text{ m}$$

$$3.132 \pm 0.026 \text{ m} \rightarrow 3.13 \text{ m}$$

$$8.274 \pm 0.010 \text{ m} \rightarrow 8.27 \text{ m}$$

$$4.865 \pm 0.018 \text{ m} \rightarrow 4.87 \text{ m}$$

$$2.335 \pm 0.009 \text{ m} \rightarrow 2.33 \text{ m}$$

The placement of the decimal point in a number representing a physical quantity is determined by the *units* that we use for the quantity. For example,

$$17 \text{ mm} = 1.7 \text{ cm} = 0.017 \text{ m} = 0.000 \ 017 \text{ km}$$

Each of these lengths is given to *two* significant figures. That is, we count significant figures from left to right, ignoring all preceding zeros.

$$2.83 \text{ g} \quad \rightarrow 3 \text{ significant figures}$$

$$0.37 \text{ kg} \ \rightarrow 2 \text{ significant figures}$$

$$0.073 \text{ m} \rightarrow 2 \text{ significant figures}$$

Because we can always express a result in many different units, we must be careful to write the result in a form that exhibits the significant figures in an unambiguous way. For example, if we have a result of 2.17 km, implying that the 7 is the last significant figure, we should not write this as 217,000 cm, which implies that there are *six* significant figures. Instead we should use the powers-of-10 notation and give the result as 2.17×10^5 cm. This procedure allows us automatically to terminate the number of digits with the last significant figure. If the first zero of 217,000 cm were significant, we would write 2.170×10^5 cm.

$$3.72 \times 10^4 \text{ g} \quad \rightarrow 3 \text{ significant figures}$$
$$6.80 \times 10^5 \text{ m} \quad \rightarrow 3 \text{ significant figures}$$
$$0.79 \times 10^{-3} \text{ cm} \quad \rightarrow 2 \text{ significant figures}$$
$$2.300 \times 10^4 \text{ s} \quad \rightarrow 4 \text{ significant figures}$$

EXERCISES

Express the following results in terms of significant figures:

1. 3.275 ± 0.002 g   (Ans.   8)

2. $(6.874 \pm 0.017) \times 10^8$ m   (Ans. 209)

3. 4.3276 ± 0.0008 kg   (Ans. 309)

4. 4.3274 ± 0.0004 kg   (Ans. 343)

5. 1.2 ± 0.7 cm   (Ans. 174)

6. 3.617 ± 0.003 km (Express in cm)   (Ans.   41)

7. 4.825 ± 0.007 kg (Express in g)   (Ans.   69)

8. If $L = 3.276 \pm 0.002$ m, L probably lies between what two values?   (Ans.   32)

9. How many significant figures are there in the result, $L = 0.00376$ km?   (Ans. 275)

Round-off each of the following to 3 significant figures:

10. 7.683 cm   (Ans. 266)

11. 3.777 m   (Ans.   31)

12. 4.68175 g   (Ans.   24)

13. 0.02684 m/s   (Ans.   5)

14. 6.2374×10^4 m   (Ans. 202)

15. 0.9865×10^{-7} cm   (Ans.   6)

2.4 MULTIPLICATION AND DIVISION OF EXPERIMENTAL NUMBERS

Many physical quantities are determined indirectly by measurements of other quantities. For example, the surface area of a rectangular piece of metal is determined by measuring the length and the width.

The velocity of an object is determined by measuring the distance moved and the time interval required for the movement. In each case we have two experimental quantities that must be combined to produce the desired result. Each of the input quantities has associated with it an experimental uncertainty (a probable error); how do we determine the uncertainty in the result?

Suppose, for example, that we wish to determine the area of a small rectangular sample of metal. Using calipers, we measure the length to be 9.8 mm and the width to be 3.2 mm. To each of these measurements we assign an uncertainty of ±0.1 mm. That is, our input quantities are

$$l = 9.8 \pm 0.1 \text{ mm (relative error} = 1\%)$$

$$w = 3.2 \pm 0.1 \text{ mm (relative error} = 3\%)$$

with relative errors of $0.1/9.8 = 0.01$ (1%) for l and $0.1/3.2 = 0.03$ (3%) for w. The area is

$$A = l \times w = (9.8 \text{ mm}) \times (3.2 \text{ mm}) = 31.36 \text{ mm}^2$$

What uncertainty do we assign to A? If the last digit of the result (i.e., 6) were significant, then the area would have an error of approximately ±0.01, or a relative error of approximately $0.01/31.36 = 0.0003$ (0.03%). Surely the area cannot be uncertain by only 0.03% when the length and width have errors of 1% and 3%, respectively! Therefore, not all the figures in the product $l \times w$ are significant. We can see this as follows:

First, we compute the area by taking values for l and w corresponding to the *lower* end of the range of probable error; call this area A^-:
$$A^- = (9.8 - 0.1 \text{ m}) \times (3.2 - 0.1 \text{ mm})$$

$$= (9.7 \text{ mm}) \times (3.1 \text{ mm}) = 30.07 \text{ mm}^2$$

Next, we compute A^+ by taking values for l and w corresponding to the *upper* end of the range of probable error:

$$A^+ = (9.8 + 0.1 \text{ mm}) \times (3.2 + 0.1 \text{ mm})$$

$$= (9.9 \text{ mm}) \times (3.3 \text{ mm}) = 32.67 \text{ mm}^2$$

The extreme spread in these results is

$$A^+ - A^- = 32.67 \text{ mm}^2 - 30.07 \text{ mm}^2 = 2.60 \text{ mm}^2$$

The value of A is 31.36 mm², as we found above, and we take for the range of uncertainty in A this extreme spread from A^- to A^+. But A lies halfway between A^- and A^+, so we can state the result as $A \pm \frac{1}{2} (A^+ - A^-)$; that is,

$$A = 31.36 \pm 1.30 \text{ mm}^2$$

or, rounding off the last decimal and calculating the relative error, we find

$$A = 31.4 \pm 1.3 \text{ mm}^2 \text{ (relative error} = 4\%)$$

We can draw the following conclusions from this example. First, the relative error in the result (4%) is *larger* than the relative error in either of the input quantities (1% and 3%). This is easy to understand— we cannot *improve* the precision by simply multiplying two quantities. Second, we notice that the relative error in the area is just the *sum* of the relative errors in the length and the width. This is, in fact, a general rule[*]:

> *Relative* error in a *product* or a *quotient* equals the *sum* of the *relative* errors in the input quantities.   (2.2)

Notice that the rule applies to *quotients* as well as to *products*. (Division, after all, is just multiplication by the reciprocal.)

Example 2.4.1

What is the area (and the uncertainty) of a circle whose radius is 5.2 ± 0.2 cm?

Although we have only one measured quantity in this case, we still have a *product* to consider:

$$A = \pi r^2 = \pi \times r \times r = \pi \times (5.2 \pm 0.2 \text{ cm}) \times (5.2 \pm 0.2 \text{ cm})$$

The relative error in the radius is $0.2/5.2 = 0.04$ (4%); therefore, the relative error in the area is $2 \times 4\% = 8\%$. (There is no error in the quantity π and this multiplicative factor does not influence the relative error in the result.) Thus,

$$A = \pi \times (5.2 \text{ cm})^2 \pm 8\% = 8.5 \pm 0.7 \text{ cm}^2$$

In this case we have rounded off the result and retained only one significant figure in the uncertainty. (Compare with the example above where we carried two significant figures and wrote ± 1.3 mm^2.) Usually, it is sufficient to round off all quantities to leave one significant figure in the uncertainty of the result.

---

[*]As stated, this rule does not agree exactly with that derived from a detailed analysis of error processes, but it is sufficiently close to be adequate for all our purposes.

Example 2.4.2

In order to determine the speed of a moving object, we measure the time required for the object to pass between two points which are 1.015 ± 0.002 m apart. The measured time interval is $(3.5 \pm 0.1) \times 10^{-3}$ s. What is the speed?

The input quantities are

$$l = 1.015 \pm 0.002 \text{ m (relative error} = 0.2\%)$$

$$t = (3.5 \pm 0.1) \times 10^{-3} \text{ s (relative error} = 3\%)$$

Therefore, calculating l/t and adding the relative errors, we find

$$v = \frac{l}{t} = \frac{1.015 \text{ m}}{3.5 \times 10^{-3} \text{ s}} = (290 \text{ m/s}) \pm 3.2\%$$

so that

$$v = 290 \pm 9 \text{ m/s}$$

In this case, one of the input quantities (the length) is known with considerably higher precision than the other quantity (the time). Therefore, the uncertainty in the result is due almost entirely to the timing error. (In fact, we could have ignored the length error without changing the final error since we retain only one significant figure in the final error.)

EXERCISES

What are the uncertainties in each of the areas determined from the following length and width measurements?

1. $l = 42.7 \pm 0.4$ m, $w = 21.3 \pm 0.4$ m   (Ans. 336)

2. $l = 3.174 \pm 0.003$ cm, $w = 2.48 \pm 0.05$ cm   (Ans.   77)

3. $l = 2.03 \pm 0.04$ m, $w = 51.6 \pm 0.2$ cm   (Ans. 144)

4. What is the uncertainty in the volume of a cube, each side of which has a length of 1.07 ± 0.02 cm?   (Ans. 124)

5. Express the area of a rectangle ($l = 80.4 \pm 0.4$ cm, $w = 21.2 \pm 0.4$ cm) in terms of significant figures only to indicate the uncertainty.   (Ans. 177)

6. A motorist drives a "measured mile" (accurate to 2%) in 80.4 ± 0.4 s. What was the speed in mi/hr? (Ans. 243)

7. An object has a mass of 1.52 ± 0.03 kg and is moving with a speed of 48.2 ± 0.2 m/s. What is the kinetic energy of the object? (Refer to Table 1.4.) (Ans. 135)

2.5 ADDITION AND SUBTRACTION OF EXPERIMENTAL NUMBERS

Suppose that we prepare a certain mixture by adding 10.2 ± 0.3 g of substance A to 304 ± 3 g of substance B. What will be the mass of the mixture and what will be the uncertainty in the result? The input quantities are

$$m_A = 10.2 \pm 0.3 \text{ g}$$

$$m_B = 304 \pm 3 \text{ g}$$

The relative error in m_A is 3%, whereas the relative error in m_B is 1%. If we were to follow the prescription for computing the uncertainty as discussed in the preceding section, we would conclude that the relative error in $M = m_A + m_B$ is 4%, giving an absolute error of

$$\text{absolute error} = 0.04 \times (10.2 \text{ g} + 204 \text{ g})$$

$$= 0.04 \times 314.2 \text{ g} = 12.6 \text{ g}$$

But this error is considerably larger than the absolute error in either m_A or m_B and, in fact, is even larger than m_A itself! This method of computing the uncertainty is surely wrong.

From the standpoint of the physical process involved in this problem, we can argue as follows. Substance B alone has a mass uncertainty of ± 3 g; adding any amount of substance A will not alter that value, and the added amount of substance A carries its own mass uncertainty to the mixture. We must conclude, therefore, that the absolute error in the mixture is equal to the *sum* of the *absolute errors* in the constituents, or ± 3.3 g in this case. The general rule is*

> *Absolute* error in an *addition* or *subtraction* equals the *sum* of the *absolute* errors in the input quantities. (2.3)

---

*Again, this rule is not strictly correct but it is entirely sufficient for our purposes.

Example 2.5.1

The distance from a point P to a point P' is determined in two steps. The individual results are $l_1 = 37.2 \pm 0.2$ m and $l_2 = 43.6 \pm 0.3$ m. What is the distance PP'.

Following the general rule, we have

$$\begin{array}{r} l_1 = 37.2 \pm 0.2 \text{ m} \\ l_2 = 43.6 \pm 0.3 \text{ m} \\ \hline PP' = 80.8 \pm 0.5 \text{ m} \end{array}$$

Example 2.5.2

Measurements are made of two masses with the results $m_1 = 48.1 \pm 0.3$ g and $m_2 = 47.6 \pm 0.2$ g. These masses are placed on a beam balance and compared. What is the mass difference between m_1 and m_2?

According to the general rule we have

$$\begin{array}{r} m_1 = 48.1 \pm 0.3 \text{ g} \\ m_2 = 47.6 \pm 0.2 \text{ g} \\ \hline \text{difference} = 0.5 \pm 0.5 \text{ g} \end{array}$$

That is, the uncertainty in the mass difference is equal to the mass difference itself. This is quite reasonable when we realize that the ranges of uncertainty for the two measurements just meet, and therefore, within the limits of precision of the two measurements, the masses could, in fact, be equal.

EXERCISES

Compute the following sums and differences and their uncertainties:

1. $(3.76 \pm 0.06 \text{ g}) + (4.876 \pm 0.003 \text{ g})$ (Ans. 286)

2. $(6.37 \pm 0.02 \text{ kg}) + (78.6 \pm 0.6 \text{ g})$ (Ans. 145)

3. $(48.3 \pm 0.4 \text{ m}) - (39.6 \pm 0.8 \text{ m})$ (Ans. 240)

4. $(17.6 \pm 0.3 \text{ m}) - (48.6 \pm 0.2 \text{ cm})$ (Ans. 315)

5. What is the perimeter of a triangle with sides of lengths $l_1 = 27.6 \pm 0.2$ cm, $l_2 = 36.8 \pm 0.3$ cm, and $l_3 = 42.7 \pm 0.3$ cm?   (Ans. 150)

6. Two masses ($m_1 = 47.3 \pm 0.8$ g and $m_2 = 52 \pm$ 1 g) are placed on opposite pans of a beam balance. Compute the mass of the weight that must be placed in the pan containing m_1 in order to balance the beam.   (Ans.   70)

2.6 COUNTING STATISTICS

Suppose that we use a Geiger counter to record the radioactive emissions from a sample of uranium ore. We count the number of detected particles for some predetermined period of time—for example, 1 minute. If we repeat the measurement many times, always keeping conditions the same, we find that we do not record exactly the same number of counts in each 1-min interval. We might obtain a series of counts such as

$$1013, 984, 1021, 1003, 995, 978, 1019, 989$$

The reason for the variation in the number of counts per minute is that there is no way to predict the instant when any given radioactive nucleus will decay. We can only specify the *probability* that a given nucleus will decay during a certain specified time interval. When we deal with large numbers of radioactive nuclei (as in a sample of uranium ore), the decay particles are emitted from the sample spaced at *random* intervals of time. Therefore, we cannot expect the same number of decay events to take place within two identical but short intervals of time. If we lengthen the time interval for counting, we again find that the number of counts varies from one measurement to the next; however, the *relative* (or percentage) variation is *smaller* than for the shorter time interval.

The general rule for the counting of radioactive decay events (or for *any* type of process that occurs with random time spacing) is the following. If a large number of measurements is made and the *average* number of counts per minute is found to be N, then the individual measurements will be distributed in such a way that approximately one-half of the individual results will lie in the range $N \pm \sqrt{N}$. That is, each individual measurement has associated with it a kind of "probable error" equal to $\sqrt{N}$.

We can immediately see that longer counting times provide increased precision in measuring the rate of occurrence of decay events. If we wish to know the decay rate to $\pm 10\%$, then we need count only long enough to record 100 counts, for then we have $100 \pm \sqrt{100} = 100 \pm 10 = 100 \pm 10\%$. However, if we wish to increase the precision by a factor of 10 to $\pm 1\%$, we must count 100 times as long in order to obtain $10{,}000 \pm \sqrt{10{,}000} = 10{,}000 \pm 100 = 10{,}000 \pm 1\%$.

Example 2.6.1

A detector records 41,376 counts from a certain radioactive sample in a period of 7.2 min. What is the decay rate of the material?

The uncertainty in the number of decays is $\sqrt{N}$, where $N = 41,376$. Therefore,

$$N \pm \sqrt{N} = 41,376 \pm \sqrt{41,376} = 41,376 \pm 203$$

and the decay rate is

$$\text{decay rate} = \frac{41,376 \pm 203}{7.2 \times 60 \text{ s}} = 95.8 \pm 0.5 \text{ s}^{-1}$$

The precision of the result is

$$\text{precision} = \frac{\sqrt{N}}{N} = \frac{1}{\sqrt{N}} = \frac{1}{203} = 0.5\%$$

EXERCISES

1. In a 1-min interval, a detector records 906 counts from a radioactive sample. For the next 1-min counting interval, the probability is 50 per cent that the number of counts will lie in what range? (Ans. 328)

2. How many counts must be recorded in order to measure a decay rate to a precision of 2%? (Ans. 2)

3. A traffic counter records 421 automobiles passing a certain point between 9 A.M. and 10 A.M. The next day, during the same hour, only 307 automobiles were recorded. What can you conclude from this observation? (Ans. 167)

4. Is there any difference (from the standpoint of the uncertainty in the final result) of taking ten 1-min counting periods or one 10-min counting period to determine an average counting rate? (Ans. 340)

CHAPTER 3

ALGEBRA

3.1 PROPERTIES OF NUMBERS

Some of the basic properties of numbers that we will be using in solving problems are summarized as follows:

1. The *real number* system* is composed of two types of numbers:
   A. *Rational numbers* — the positive and negative integers and zero ($\ldots,-3,-2,-1,0,1,2,3,\ldots$); and the fractions m/n where m and n are integers.
   B. *Irrational numbers* — numbers expressible as infinite decimals e.g., $\pi = 3.14159\ldots$ or $\sqrt{2} = 1.41421\ldots$) but not as ratios of integers. (Thus, $2/3 = 0.66666\ldots$ is a *rational* number.)
2. All real numbers obey the following rules:

$$a + b = b + a \tag{3.1}$$

$$(a + b) + c = a + (b + c) \tag{3.2}$$

$$ax + bx = (a + b)x \tag{3.3}$$

We will be continually using these properties but we will not refer each time to a specific property or rule.

3.2 SOLVING SIMPLE EQUATIONS

An *equation* expresses the equality between two quantities or combinations of quantities; for example,

$$A + B = C + D$$

Equations can be manipulated in various ways without affecting the validity of the equality. These operations are of the following types:

(a) Addition (or subtraction) of the same quantity to (or from) both sides of an equation:

---

*Brief mention of *imaginary* numbers is made in Section 3.4.

$$(A + B) + M = (C + D) + M$$

$$(A + B) - N = (C + D) - N$$

(b) Multiplication (or division) by the same quantity of *both* sides of an equation:

$$(A + B) \times M = (C + D) \times M$$

$$\frac{(A + B)}{N} = \frac{(C + D)}{N}$$

(c) Raising of *both* sides of an equation to the same power or extracting the same root of *both* sides of an equation:

$$(A + B)^3 = (C + D)^3$$

$$\sqrt{A + B} = \sqrt{C + D}$$

In general, if the same mathematical operation is applied to both sides of an equation, the result is a valid equation.

We will consider in this section equations which can be reduced to the form

$$ax + b = 0 \tag{3.4}$$

This is a *linear* equation, that is, an equation for an unknown quantity x in which x appears only to the first power. (Equations in which x^2 appears are *quadratic* equations and will be discussed in Section 3.4.)

The basic operations described above are sufficient to solve all simple linear equations. The object is to select the proper operations to isolate the unknown quantity x.

Example 3.2.1
Solve $3x - 7 = 0$:

Step 1. Add 7 to both sides:

$$(3x - 7) + 7 = 7$$

$$3x = 7$$

Step 2. Divide both sides by 3:

$$\frac{3x}{3} = \frac{7}{3}$$

$$x = \frac{7}{3}$$

Example 3.2.2
Solve $ax - b = 4x - 7$

Step 1. Subtract $(4x - b)$ from both sides:

$$(ax - b) - (4x - b) = (4x - 7) - (4x - b)$$
$$ax - 4x = b - 7$$
$$(a - 4)x = b - 7$$

Step 2. Divide both sides by $(a - 4)$:

$$\frac{(a - 4)x}{a - 4} = \frac{b - 7}{a - 4}$$
$$x = \frac{b - 7}{a - 4}$$

Example 3.2.3
Solve $\dfrac{1}{1 - x} = a$

Step 1. Multiply both sides by $(1 - x)$:

$$\frac{1 - x}{1 - x} = a(1 - x)$$
$$1 = a - ax$$

Step 2. Add $(ax - 1)$ to both sides:

$$1 + (ax - 1) = a - ax + (ax - 1)$$
$$ax = a - 1$$

Step 3. Divide both sides by a:

$$x = \frac{a - 1}{a}$$

Example 3.2.4

Solve $\dfrac{a}{by^{\frac{1}{3}}} = c^{\frac{2}{3}}$

Step 1. Cube both sides:

$$\frac{a^3}{b^3 y} = c^2$$

Step 2. Multiply both sides by $\dfrac{y}{c^2}$:

$$\frac{a^3}{b^3 y} \times \frac{y}{c^2} = c^2 \times \frac{y}{c^2}$$

$$\frac{a^3}{b^3 c^2} = y$$

Example 3.2.5

An automobile travels at a speed of 40 miles per hour. How much time is required to go 340 miles?

The basic equation is

$$\text{distance} = \text{velocity} \times \text{time}$$

In symbols this becomes

$$x = vt \tag{1}$$

The unknown quantity is the time t, and solving for t,

$$t = \frac{x}{v} = \frac{340 \text{ mi}}{40 \text{ mi/hr}} = 8.5 \text{ hr} \tag{2}$$

Notice the way in which the units are manipulated:

$$\frac{\text{mi}}{\frac{\text{mi}}{\text{hr}}} = \frac{\cancel{\text{mi}}}{\cancel{\text{mi}}} \times \frac{\text{hr}}{\cancel{\text{hr}}} = \text{hr} \tag{3}$$

EXERCISES

In each case, solve for x:

1. $3x + 8 = 50 - 4x$ (Ans. 235)

2. $12x - a = b + 7 + 4x$ (Ans. 247)

2. $\dfrac{3}{2}x + 12 = \dfrac{7}{2}x - 4$ (Ans. 126)

4. $\dfrac{a}{x+b} = c$ (Ans. 81)

5. $\dfrac{3}{x-4} = \dfrac{7}{x-5}$ (Ans. 103)

3.3 SIMULTANEOUS LINEAR EQUATIONS

In certain types of physical problems we find that we have an equation involving more than a single unknown. For example,

$$6x + 2y = 6$$

Such an equation does not yield a unique solution; it is easy to verify that $(x = 1,\ y = 0)$, $(x = 4,\ y = -9)$, and $(x = 1/2,\ y = 3/2)$ are all solutions of the above equation.

In order to have a unique solution for a problem involving *two* unknowns, we must have *two* equations. In general, if we have a problem with n unknown quantities, we need n equations in order to effect a solution.

Two methods of solving a pair of simple linear equations are given in the following example.

Example 3.3.1

Consider the pair of equations,

$$6x + 2y = 6 \tag{1}$$

$$8x - 4y = 28 \tag{2}$$

The solution can be obtained by first solving one of the equations for y and then substituting this expression into the other equation. The result will be a single equation in the unknown x and the solution can be obtained by one of the methods described in the preceding section.

Solving (1) for y,

$$2y = 6 - 6x$$
$$y = 3 - 3x \tag{3}$$

Substituting (3) into (2), we find

$$8x - 4(3 - 3x) = 28$$
$$8x - 12 + 12x = 28$$
$$20x = 40$$
$$x = 2 \tag{4}$$

Finally, substituting (4) into (1), we obtain

$$6 \cdot 2 + 2y = 6$$
$$2y = -6$$
$$y = -3 \tag{5}$$

Therefore, the solution is $(x = 2, y = -3)$.

Alternatively, we can solve these equations by multiplying (1) by 2 and then adding to (2):

$$2 \cdot (6x + 2y) = 2 \cdot 6$$
$$12x + 4y = 12 \tag{6}$$

Then, adding (2) and (6),

$$8x - 4y = 28$$
$$\underline{12x + 4y = 12}$$
$$20x \qquad = 40$$
$$x = 2 \tag{7}$$

And substituting (7) into (1) again produces $y = -3$.

Notice that we choose to multiply (1) by 2 (and not by some other number) in order to make the term involving y equal in magnitude to the corresponding term in (2). Then, when the two equations are added, the y-term cancels. In some cases, we must multiply each equation by a different factor in order to ensure that one of the terms will cancel. Thus, we could have multiplied (1) by 4 and (2) by 3 and then *subtracted* the resulting equations to eliminate the x-terms. (Work out the example in this way.)

EXERCISES

Solve the following pairs of equations:

1. $x + y = 4$
   $x - y = 8$ (Ans.   59)

2. $4x + 3y = 8$
    $12x - \ y = 4$ (Ans. 295)

3. $5x + 2y = 38$
   $3y - 2x = 19$ (Ans. 186)

3.4 QUADRATIC EQUATIONS

Quadratic equations involve the *second* power of the unknown quantity and can be reduced to the form

$$ax^2 + bx + c = 0 \qquad (3.5)$$

Sometimes $b = 0$ and then we have a particularly simple equation,

$$ax^2 + c = 0 \qquad (3.5a)$$

This equation can be reduced to a linear equation by subtracting c from both sides, dividing both sides by a, and then taking the square root of both sides:

$$ax^2 = -c$$

$$x^2 = -\frac{c}{a}$$

$$x = \pm \sqrt{-\frac{c}{a}}$$

Notice that this solution for x involves the square root of a quantity that carries a negative sign. However, if $a = 2$, $c = -8$, then

$$x = \pm \sqrt{-\frac{c}{a}} = \pm \sqrt{-\frac{(-8)}{2}} = \pm \sqrt{4} = \pm 2$$

and there is no difficulty because the two negative signs combine to produce a positive sign.

On the other hand, if $a = 2$, $c = 8$, then

$$x = \pm \sqrt{-\frac{c}{a}} = \pm \sqrt{-\frac{8}{2}} = \pm \sqrt{-4}$$

We are now faced with the problem of finding a number which, when multiplied by itself, produces -4. *There is no such real number.* We must go outside the realm of real numbers to find a solution to this problem.

Notice that we can always express the square root of a negative number in the following way:

$$\sqrt{-4} = \sqrt{(-1)(4)} = \sqrt{-1} \times \sqrt{4} = \sqrt{-1} \times (\pm 2)$$

The quantity $\sqrt{-1}$ is called an *imaginary* quantity and is denoted by the symbol i:

$$\sqrt{-1} = i; \quad i^2 = -1 \tag{3.6}$$

Therefore, we can write the solution to the problem as

$$x = \pm \sqrt{-4} = \pm 2i$$

Imaginary quantities appear in the mathematical descriptions of certain types of physical problems, but we will not pursue such problems here and will not discuss imaginary quantities further.

Returning to the general quadratic equation, Equation 3.5, we will state (but not derive) the solution. The solution of the equation

$$ax^2 + bx + c = 0 \tag{3.5}$$

is

$$x = \frac{-b \pm \sqrt{b^2 - 4ac}}{2a} \tag{3.7}$$

In the event that $b^2 \geq 4ac$, the result will be a *real* quantity. Before solving a quadratic equation, always make this test. If $b^2 = 4ac$, the square-root term vanishes and we have a unique result for x. If $b^2 > 4ac$, there will be *two* possible results for x, depending on which sign we take for the square root. Each of these results is a true solution of the equation and each is equally valid.

Example 3.4.1

Solve the equation

$$x^2 + 4x + 4 = 0 \tag{1}$$

We identify

$$a = 1, \; b = 4, \; c = 4 \tag{2}$$

and we note that $b^2 = 4ac$. Then, using Equation 3.7, we have

$$x = \frac{-4 \pm \sqrt{4^2 - 4 \cdot 1 \cdot 4}}{2 \times 1} = \frac{-4 \pm \sqrt{16 - 16}}{2} \tag{3}$$

Since the square root term is zero, we find

$$x = -2 \tag{4}$$

The answer is readily verified to be correct by substitution of (4) into (1).

Example 3.4.2

Solve the equation

$$7x^2 - 8x + 1 = 0 \tag{1}$$

We identify

$$a = 7, \; b = -8, \; c = 1 \tag{2}$$

and we see that $b^2 = 64$ and $4ac = 28$, so that $b^2 > 4ac$. Using Equation 3.7, the solution is

$$x = \frac{-(-8) \pm \sqrt{(-8)^2 - 4 \cdot 7 \cdot 1}}{2 \times 7}$$

$$= \frac{8 \pm \sqrt{64 - 28}}{14} = \frac{8 \pm \sqrt{36}}{14}$$

$$= \frac{8 \pm 6}{14} \tag{3}$$

There are now two possible solutions, one corresponding to

each of the signs of the square-root term. We label these x_+ and x_-:

$$x_+ = \frac{8+6}{14} = 1 \qquad (4)$$

$$x_- = \frac{8-6}{14} = \frac{2}{14} = \frac{1}{7} \qquad (5)$$

Again, it is easy to verify by substitution into (1) that each solution is valid.

In the example above there were *two* possible solutions. In a physics problem, one can usually determine which solution is physically meaningful by examining the *physics* of the situation. (Sometimes *both* solutions are valid.) The following example demonstrates such a procedure.

Example 3.4.3

The position x of a particle at a time t, moving in either direction along a straight line, is given by

$$x = x_0 + v_0 t + \frac{1}{2} at^2 \qquad (1)$$

where x_0 is the position at time $t = 0$ and v_0 is the velocity at $t = 0$, and where a is the acceleration of the particle.

Suppose that a ball is thrown vertically downward with a velocity of 60 ft/s from the top of a 100-ft building. If the ball is released at $t = 0$, when will it strike the ground?

If we measure x from the top of the building *downward*, then $x_0 = 0$, $x = 100$ ft, and $v_0 = 60$ ft/s. The acceleration experienced by the particle is g, the acceleration due to gravity, and has the value $g = 32$ ft/s² (see Eq. 1.21).

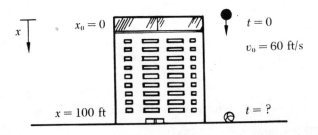

Writing (1) in standard form with $a = g$ and $x_0 = 0$, we have

$$\frac{1}{2}gt^2 + v_0 t - x = 0 \tag{2}$$

Solving for t,

$$t = \frac{-v_0 \pm \sqrt{v_0^2 - (4)\left(\frac{1}{2}g\right)(-x)}}{g} \tag{3}$$

Substituting the values for v_0, g, and x, we find

$$t = \frac{-60 \pm \sqrt{(60)^2 - (4)(16)(-100)}}{32}$$

$$= \frac{-60 \pm \sqrt{3600 + 6400}}{32}$$

$$= \frac{-60 \pm \sqrt{10^4}}{32}$$

$$= \frac{-60 \pm 100}{32} \tag{4}$$

The two possible solutions are

$$t_+ = \frac{-60 + 100}{32} = \frac{40}{32} = 1.25 \text{ s} \tag{5}$$

$$t_- = \frac{-60 - 100}{32} = -\frac{160}{32} = -5 \text{ s} \tag{6}$$

Clearly, the solution $t_- = -5$ s has no meaning in this case because the ball was released at $t = 0$. The proper solution is therefore $t_+ = 1.25$ s.

EXERCISES

Solve the following equations:

1. $3x^2 - 6x - 9 = 0$ (Ans. 359)

2. $5x^2 + 2x - 24 = 0$ (Ans. 194)

3. $x^2 - 9x + 20 = 0$ (Ans. 253)

4. $3x^2 + 8x - \frac{19}{4} = 0$ (Ans. 179)

5. $x = x_0 + v_0 t + \frac{1}{2}gt^2$, where $x_0 = 100$ ft, $x = 0$,

   $v_0 = -60$ ft/s, and $g = -32$ ft/s^2 (Ans. 268)

3.5 THE BINOMIAL EXPANSION

The multiplication of two algebraic quantities proceeds in the following way:

$$(a + b)(c + d) = a(c + d) + b(c + d)$$
$$= ac + ad + bc + bd$$

$$(a + b)^2 = (a + b)(a + b)$$
$$= a(a + b) + b(a + b)$$
$$= a^2 + ab + ba + b^2$$
$$= a^2 + 2ab + b^2$$

$$(a + b)^3 = (a + b)(a + b)^2$$
$$= (a + b)(a^2 + 2ab + b^2)$$
$$= a(a^2 + 2ab + b^2) + b(a^2 + 2ab + b^2)$$
$$= a^3 + 2a^2b + ab^2 + a^2b + 2ab^2 + b^3$$
$$= a^3 + 3a^2b + 3ab^2 + b^3$$

The general rule for the nth power of an algebraic quantity is given by the binomial theorem:

$$(a + b)^n = a^n + na^{n-1}b + \frac{n(n-1)}{1 \cdot 2} a^{n-2}b^2$$
$$+ \frac{n(n-1)(n-2)}{1 \cdot 2 \cdot 3} a^{n-3}b^3 + \ldots \tag{3.8}$$

Notice that when n is a positive integer this series does not continue indefinitely but terminates after $n + 1$ terms.

This expression permits writing the expansion of $(a + b)^3$ much more rapidly than does the method used above:

$$(a + b)^3 = a^3 + 3a^2b + \frac{3 \cdot 2}{1 \cdot 2} ab^2 + \frac{3 \cdot 2 \cdot 1}{1 \cdot 2 \cdot 3} a^0b^3$$
$$= a^3 + 3a^2b + 3ab^2 + b^3$$

We find many uses for the binomial expansion in physics problems. One of the most common cases is that in which $a = 1$. Then, Equation 3.8 becomes

$$(1 + b)^n = 1 + nb + \frac{n(n-1)}{1 \cdot 2} b^2 + \frac{n(n-1)(n-2)}{1 \cdot 2 \cdot 3} b^3 + \ldots \tag{3.9}$$

For example,

$$(1+b)^3 = 1 + 3b + 3b^2 + b^3$$

We can always convert an expression of the type $(a+b)^n$ to one of the form $(1+b)^n$ by *factoring* the original expression. For example,

$$(3+x)^3 = 3^3\left(1+\frac{x}{3}\right)^3 = 27(1+y)^3$$

where we have used y to represent $x/3$.

Reciprocals can be obtained by using negative values of n:

$$(1+b)^{-1} = 1 - b + b^2 - b^3 + \cdots \tag{3.10}$$

$$(1+b)^{-2} = 1 - 2b + 3b^2 - 4b^3 + \cdots \tag{3.11}$$

In these cases, notice that the series of terms does not *terminate* as it does for positive integer values of n. That is, the expansion for $(1+b)^{-1}$ or for $(1+b)^{-2}$ consists of an *infinite series* of terms. Unless $|b| < 1$, each term in the series is larger than the preceding term and the series is of no value in evaluating the quantity. We say that the series *converges* only for $|b| < 1$.

Example 3.5.1

Obtain the series expansion for $(4+x)^{-2}$.

First, we factor 4 out of the expression and substitute y for $x/4$:

$$(4+x)^{-2} = (4)^{-2}\left(1+\frac{x}{4}\right)^{-2} = \frac{1}{16}(1+y)^{-2}$$

Then, using Equation 3.11,

$$\frac{1}{16}(1+y)^{-2} = \frac{1}{16}(1 - 2y + 3y^2 - 4y^3 + \cdots)$$

If $y = 2$ (so that $x = 4y = 8$), we find

$$(4+x)^{-2} = \frac{1}{16}(1 - 2\cdot2 + 3\cdot4 - 4\cdot8 + \cdots)$$

$$= \frac{1}{16}(1 - 4 + 12 - 32 + \cdots)$$

and the value of the series fluctuates widely as we add more and more terms. This series does not converge. On the other hand, if we have $y = 0.1$ (so that $x = 4y = 0.4$), then we find

$$(4 + x)^{-2} = \frac{1}{16} (1 - 2 \cdot 0.1 + 3 \cdot 0.01 - 4 \cdot 0.001 + \cdots)$$

$$= \frac{1}{16} (1 - 0.2 + 0.03 - 0.004 + \cdots)$$

$$\cong \frac{1}{16} (0.826) = 0.05162$$

Adding more terms will improve the result in the next decimal place, but the result will never differ greatly from 0.05162.

Expansions for the *roots* of quantities can be obtained by substituting the appropriate fractional value for n:

$$(1 + b)^{\frac{1}{2}} = 1 + \frac{1}{2} b - \frac{1}{8} b^2 + \frac{1}{16} b^3 - \cdots \tag{3.12}$$

$$(1 - b)^{-\frac{1}{2}} = 1 - \frac{1}{2} b + \frac{3}{8} b^2 - \frac{5}{16} b^3 + \cdots \tag{3.13}$$

(Verify these expansions.) Again, we must have $|b| < 1$ for convergence.

One of the important applications of Equations 3.12 and 3.13 is in the theory of relativity where we are frequently called upon to evaluate $\sqrt{1 - \beta^2}$ and $1/\sqrt{1 - \beta^2}$, where β is the ratio of the speed of a particle to the speed of light, $\beta = v/c$. Substituting $-\beta^2$ for b, we find

$$\sqrt{1 - \beta^2} = 1 - \frac{1}{2} \beta^2 - \frac{1}{8} \beta^4 - \frac{1}{16} \beta^6 - \cdots \tag{3.14}$$

$$\frac{1}{\sqrt{1 - \beta^2}} = 1 + \frac{1}{2} \beta^2 + \frac{3}{8} \beta^4 + \frac{5}{16} \beta^6 + \cdots \tag{3.15}$$

If the speed of the particle is much smaller than the speed of light, $v \ll c$, then $\beta^2 \ll 1$. For example, if $v = 3 \times 10^7$ m/s, we have (since $c = 3 \times 10^8$ m/s)

$$\beta = \frac{v}{c} = \frac{3 \times 10^7 \text{ m/s}}{3 \times 10^8 \text{ m/s}} = 10^{-1}$$

so that

$$\beta^2 = (10^{-1})^2 = 10^{-2}$$

$$\beta^4 = (10^{-1})^4 = 10^{-4}$$

$$\beta^6 = (10^{-1})^6 = 10^{-6}$$

Then, Equation 3.14 becomes

$$\sqrt{1 - \beta^2} = 1 - \frac{1}{2} \times 10^{-2} - \frac{1}{8} \times 10^{-4} - \frac{1}{16} \times 10^{-6} - \cdots$$

$$= 1 - 0.005 - 0.0000125 - 0.0000000625 - \cdots$$

$$= 0.9949874375 \cong 0.995$$

It is apparent that the terms after $-\frac{1}{2}\beta^2$ do not influence the results to any significant extent. Therefore, if $\beta^2 \ll 1$, for all practical purposes we can use the approximate expressions,

$$\sqrt{1 - \beta^2} \cong 1 - \frac{1}{2}\beta^2 \left.\right\} \quad \beta^2 \ll 1 \tag{3.16}$$

$$\frac{1}{\sqrt{1 - \beta^2}} \cong 1 + \frac{1}{2}\beta^2 \tag{3.17}$$

Example 3.5.2

According to relativity theory, the mass of a particle depends on its speed according to the equation

$$m = \frac{m_0}{\sqrt{1 - \beta^2}} \tag{1}$$

where m_0 is the mass of the particle at rest. What is the mass of a hydrogen atom which is moving with a speed of 10^8 m/s?

According to the result of Exercise 2 in Section 1.4, the mass of a hydrogen atom (at rest) is

$$m_0 = 1.67 \times 10^{-24} \text{ g} \tag{2}$$

We also have

$$\beta = \frac{v}{c} = \frac{10^8 \text{ m/s}}{3 \times 10^8 \text{ m/s}} = 3.3 \times 10^{-1} \tag{3}$$

Using Equation 3.17, we can write

$$m \cong m_0 \left(1 + \frac{1}{2}\beta^2\right)$$

$$= (1.67 \times 10^{-24} \text{ g}) \times [1 + \frac{1}{2}(3.3 \times 10^{-1})^2]$$

$$= (1.67 \times 10^{-24} \text{ g}) \times (1 + 0.05)$$

$$= 1.75 \times 10^{-24} \text{ g}$$

That is, the mass of the moving hydrogen atom is 1.05 times (or 5% greater than) the mass of a hydrogen atom at rest.

We have used the binomial expansion to enable us to make *approximate* calculations of certain quantities. There is nothing "wrong" in obtaining an approximate answer to a problem. In many cases, the approximate answer is quite sufficient for all practical purposes. For example, suppose that we wish to compute the time delay between the instant that an Apollo astronaut transmits a radio signal from the moon and the instant when that signal is received by Mission Control in Houston. The time delay is

$$\text{Time delay} = \frac{\text{distance from Moon to Earth}}{\text{speed of radio signal}}$$

The speed of the radio signal is just the speed of any electromagnetic signal, namely, the speed of light. We know this speed to a precision of 100 m/s, so it requires *seven* significant figures to express this speed (see Example 2.2.1). Similarly, recent experiments in which laser signals from the Earth are reflected from mirrors on the moon have increased our knowledge concerning the Earth-Moon distance to a comparable precision (0.3 ppm). Therefore, to compute the time delay would require the division of one seven-digit number by another seven-digit number, producing an answer that is precise to seven significant figures. To what practical use could we put such a number? Absolutely none! So we might as well have computed the time delay using *approximate* values for the distance and the speed:

$$\text{Time Delay} = \frac{3.84 \times 10^8 \text{ m}}{3.00 \times 10^8 \text{ m/s}} \doteq 1.28 \text{ s}$$

Surely this is as precise an answer as we would need for almost every purpose.

Even in pursuing a *precise* science, such as physics, there is nothing disreputable in making approximate calculations. Remember—"That which is good enough is best!"

EXERCISES

1. Expand $(1 + x)^{\frac{3}{2}}$. (Ans. 276)

2. Expand $(1 + x)^{-\frac{3}{2}}$. (Ans. 305)

3. Examine Equation 3.9 and deduce the next term in the expansion. (Ans. 76)

4. What is the approximate value of $\sqrt{1.018}$? (Ans. 356)

5. If $x \ll 1$, what is the approximate value of
$\frac{1}{(1 - x)}$? (Ans. 267)

6. If $x \ll 1$, what is the approximate value of
$(1 + x)^{\frac{1}{3}}$? (Ans. 215)

7. According to relativity theory, the length l of an object in motion is related to the length l_0 of the same object when at rest by $l = l_0 \sqrt{1 - \beta^2}$. If a meter stick moves with a velocity $v = 2 \times 10^7$ m/s, what will be its length?

(Ans. 153)

3.6 SUMMATIONS

In many types of physical problems we find it necessary to add a series of related numbers. For example, in order to determine the total mass M of a system composed of many individual masses, $m_1, m_2, m_3, m_4, \ldots$, we would write

$$M = m_1 + m_2 + m_3 + m_4 + \cdots + m_N \tag{3.18}$$

where N represents the total number of particles in the system. Because this situation arises so frequently, a special notation has been devised to abbreviate lengthy summations. In order to represent Equation 3.18, we write

$$M = \sum_{i=1}^{N} m_i \tag{3.19}$$

Here, we have written a typical particle mass as m_i and we allow i (which is called the *summation index*) to take on the integer numbers from 1 to N. That is, m_i (for $i = 1$) $= m_1$, m_i (for $i = 2$) $= m_2$, etc. The symbol Σ (which is called the *summation symbol*) means "sum all of the quantities which follow." Beneath Σ is the indication of the index to be summed (here, i) and the *starting* value of the index (here, 1); above the Σ is the *terminating* value of the index (here, N).

If we sum the integers from 1 to N, the result is $\frac{1}{2} N(N+1)$. In the notation of Equation 3.19, this is expressed as

$$\sum_{i=1}^{N} i = \frac{1}{2} N(N+1) \tag{3.20}$$

(Try this for a few values of N.) Or, if we sum the *squares* of the integers from 1 to N, we find

$$\sum_{m=1}^{N} m^2 = \frac{N}{6} (2N^2 + 3N + 1) \tag{3.21}$$

Here, we have used a summation index of m instead of i; it does not matter at all what letter or symbol we use for the index because it does not appear in the final result.

Example 3.6.1

A series of masses m_i are located at various points along a straight line as indicated in the figure. Calculate the position of the *center of mass* of the system.

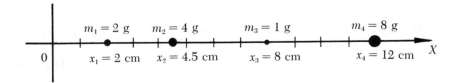

The prescription for locating the position of the center of mass is

$$x_{C.M.} = \frac{\sum\limits_{i=1}^{4} m_i x_i}{\sum\limits_{i=1}^{4} m_i}$$

First, we calculate the sum in the denominator, which is just the total mass (see Eq. 3.19):

$$M = \sum_{i=1}^{4} m_i = m_1 + m_2 + m_3 + m_4$$

$$= 2\,\mathrm{g} + 4\,\mathrm{g} + 1\,\mathrm{g} + 8\,\mathrm{g} = 15\,\mathrm{g}$$

Next, we calculate

$$\sum_{i=1}^{4} m_i x_i = m_1 x_1 + m_2 x_2 + m_3 x_3 + m_4 x_4$$

$$= [(2 \cdot 2) + (4 \cdot 4.5) + (1 \cdot 8) + (8 \cdot 12)]\ \mathrm{g\text{-}cm}$$

$$= [4 + 18 + 8 + 96]\ \mathrm{g\text{-}cm}$$

$$= 126\ \mathrm{g\text{-}cm}$$

Then,

$$x_{C.M.} = \frac{126\ \mathrm{g\text{-}cm}}{15\ \mathrm{g}} = 8.4\ \mathrm{cm}$$

and the center of mass of the system is located just to the right of m_3.

EXERCISES

1. Write out the following sum

$$\sum_{p=1}^{N} (2p-1)$$

for $N=3$ and for $N=5$. Verify that the sum is equal to N^2.

2. Write out the first few terms in the sum

$$\sum_{m=1}^{N} \frac{(-1)^m}{m}$$

(Ans. 99)

3. The following pairs of numbers (m_i, y_i) represent the masses (in grams) and the positions (in centimeters) of 4 particles that lie on the Y axis: $(3,-2)$, $(5,0)$, $(2,5)$, and $(6,10)$. Compute the center of mass of the system.

(Ans. 19)

GEOMETRY

4.1 PROPERTIES OF SIMPLE GEOMETRIC FORMS

When working physics problems it is frequently necessary to know the properties of certain simple geometric forms. The most commonly encountered forms are rectangles, parallelepipeds, triangles, circles, spheres, and cylinders, and the most important properties are the areas and the volumes. Table 4.1 summarizes these properties for these six simple forms.

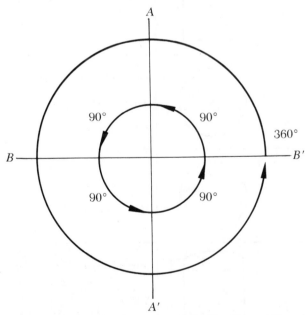

FIGURE 4.1 A complete circle contains 360°. The lines AA' and BB' are perpendicular.

The specification of an *angle* is in terms of the number of *degrees of angle* (or, simply, the *number of degrees*) between the two straight lines that define the angle. If two lines are *perpendicular,* the angle formed by the lines is 90°. A complete circle (four adjacent 90° angles) contains 360° (see Fig. 4.1). A straight line can be considered to be a 180° angle. Angles less than 90° are said to be *acute*; angles greater than 90° are said to be *obtuse*.

TABLE 4.1 PROPERTIES OF SOME SIMPLE GEOMETRIC FORMS

Form	Property
Rectangle:	Area $= ab$
Parallelepiped:	Volume $= abc$
Right Triangle:	Area $= \frac{1}{2} ab$
Circle:	Circumference $= 2\pi r$ Area $= \pi r^2$ Diameter $= d = 2r$
Sphere:	Surface area $= 4\pi r^2$ Volume $= \frac{4}{3} \pi r^3$
Cylinder:	Surface area $= 2\pi r^2 + 2\pi rh$ $= 2\pi r\,(r + h)$ Volume $= \pi r^2 h$

The sum of the interior angles of any triangle is 180°, as indicated in Figure 4.2.

Any triangle that has two sides of equal length is called an *isosceles* triangle; if all three sides are of equal length, the triangle is *equilateral* (see Fig. 4.3).

If two triangles have two sides and the included angle (the angle between these sides) equal, then the triangles are equivalent and are said to be *congruent*. Similarly, if the triangles have two angles and the included side equal, they are also congruent (see Fig. 4.4).*

---

*Notice that two congruent triangles need not be *identical*; one could be the mirror image of the other. (Test this statement.)

FIGURE 4.2 The interior angles of any triangle sum to 180°.

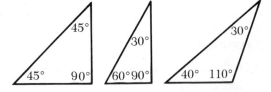

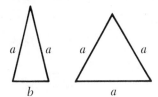

FIGURE 4.3 An *isosceles* and an *equilateral* triangle.

FIGURE 4.4 Two pairs of *congruent* triangles.

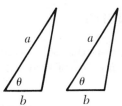

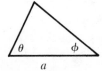
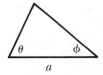

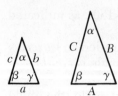

FIGURE 4.5 Two *similar* triangles.

Two triangles are said to be *similar* if each has identical interior angles (see Fig. 4.5). Similar triangles have the same *shape,* but they need not be congruent. Similar triangles have the important property that the ratios of the lengths of their sides are equal; thus, for the triangles in Figure 4.5,

$$\frac{a}{b} = \frac{A}{B}; \quad \frac{a}{c} = \frac{A}{C}; \quad \frac{a}{A} = \frac{c}{C}; \quad \text{etc.} \tag{4.1}$$

EXERCISES

1. What is the volume of a cube with sides of length 5 cm?   (Ans. 80)

2. A cube has a surface area of 24 cm². What is its volume?   (Ans. 214)

3. What is the area of a circle whose diameter is 8 cm?   (Ans. 185)

4. What is the volume of a sphere whose radius is $10^{\frac{1}{3}}$ cm?   (Ans. 348)

5. A cylinder is 4 cm high and has a radius of 2 cm. What is the volume?   (Ans. 342)

6. The radius of the Moon is 1.74×10^8 cm. What is the volume of the Moon? (Remember significant figures!)   (Ans. 187)

4.2 DENSITY

If we know the volume and the mass of a substance, then we can define the *density* of that substance:

$$\text{density} = \frac{\text{mass}}{\text{volume}} \tag{4.2}$$

or, in symbols,

$$\rho = \frac{M}{V} \tag{4.3}$$

TABLE 4.2 DENSITIES OF SOME COMMON SUBSTANCES

Substance	Density (g/cm³)
Air (normal conditions*)	1.3×10^{-3}
Water	1
Aluminum	2.7
Iron	7.9
Lead	11.3
Gold	19.3

*Temperature $= 0°$ C and pressure $= 1$ atmosphere.

The density is an *intrinsic* property of a substance and does not depend on its shape or its volume. The density of water, for example, is 1 gram per cubic centimeter (1 g/cm³). (In fact, the *gram* was originally defined to be the mass of 1 cm³ of water.) The densities of some common substances are given in Table 4.2.

Example 4.2.1

Lead bricks (used for shielding radioactive materials) commonly have dimensions of 2 in × 4 in × 8 in. What is the mass of such a brick?

$$\text{Volume} = V = (2 \text{ in}) \times (4 \text{ in}) \times (8 \text{ in}) = 64 \text{ in}^3$$

$$= (64 \text{ in}^3) \times \left(\frac{2.54 \text{ cm}}{1 \text{ in}}\right)^3$$

$$= (64 \text{ in}^3) \times \left(\frac{16.39 \text{ cm}^3}{1 \text{ in}^3}\right)$$

$$= 1049 \text{ cm}^3 \tag{1}$$

Using $\rho = 11.3$ g/cm³ from Table 4.2, we have

$$M = \rho V = \left(\frac{11.3 \text{ g}}{\text{cm}^3}\right) \times (1049 \text{ cm}^3)$$
$$= 1.185 \times 10^4 \text{ g}$$
$$= 11.85 \text{ kg} \tag{2}$$

or, in British units,

$$M = (1.185 \times 10^4 \text{ g}) \times \left(\frac{1 \text{ lb}}{453.59 \text{ g}}\right)$$

$$= 26.12 \text{ lb}$$

The rule-of-thumb is that the mass of a standard lead brick is ~ 25 lb.

Example 4.2.2

What is the average density of the Earth?
The radius of the Earth is $R_E = 6.38 \times 10^8$ cm, so that

$$V = \frac{4}{3}\pi R_E^3 = \frac{4}{3}\pi(6.38 \times 10^8 \text{ cm})^3$$

$$= 1.09 \times 10^{27} \text{ cm}^3$$

The mass of the Earth is $M_E = 5.98 \times 10^{27}$ g; thus,

$$\rho = \frac{M_E}{V} = \frac{5.98 \times 10^{27} \text{ g}}{1.09 \times 10^{27} \text{ cm}^3} = 5.49 \text{ g/cm}^3$$

This is the density averaged throughout the entire Earth. Actually, the core has a considerably higher density (~ 12 g/cm³) and the mantle (the rocky material near the surface) has a density of ~ 3 g/cm³.

EXERCISES

1. What is the mass of a cube of iron with sides of length 2 cm?   (Ans. 270)

2. What is the mass of a cubic meter of air at normal conditions?   (Ans. 74)

3. A sphere of aluminum has a radius of 10 cm. What is the mass?   (Ans. 191)

4. A pipe with a diameter of 10 cm contains a column of water 10 m high. What is the mass of the water?   (Ans. 130)

5. What is the *difference* in mass between two spheres (each with radius = 5 cm), one of which is lead and one of which is gold?   (Ans. 71)

6. What is the density (the *average* density) of the Sun? (Radius = 6.96×10^{10} cm, mass = 1.99×10^{33} g.)   (Ans. 92)

4.3   RECTANGULAR COORDINATE SYSTEMS

A *map* is a device for determining the position of a place on the surface of the Earth. Although the Earth is spherical, if we confine our attention to a relatively small region, we can consider the region to be *flat*. Such a map, therefore, represents the positions of places on a

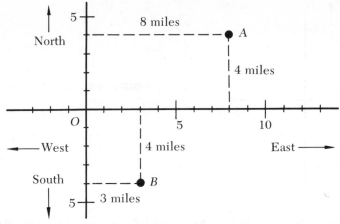

FIGURE 4.6 The location of two points on a rectangular mileage map.

horizontal *plane.* By specifying the north-south position (the *latitude*) and the east-west position (the *longitude*) we can uniquely locate a particular place; the latitude and the longitude are the *coordinates* of the position.

Instead of measuring the coordinates of a point in terms of *degrees* of latitude and *degrees* of longitude, let us simplify the situation and specify the coordinates by stating the north-south and the east-west distances from a central point called the *origin.* Figure 4.6 shows this method for locating the position of a point *A* relative to the origin *O*: *A* is located 8 miles east of *O* and 4 miles north of *O*. Also shown is a point *B* that is 3 miles east of *O* and 4 miles south of *O*.

It is not necessary that the mileage scales along the north-south and east-west directions be the same. (Of course, if the scales are *not* the same, we will have a distorted picture of the location of the points, but nevertheless the points will be uniquely and correctly represented.) Figure 4.7 shows the same points *A* and *B* on a map with different north-south and east-west scales.

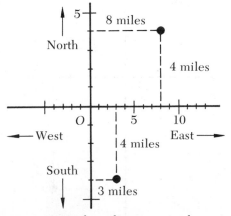

FIGURE 4.7 The same rectangular mileage map as shown in Figure 4.6 but with different north-south and east-west mileage scales.

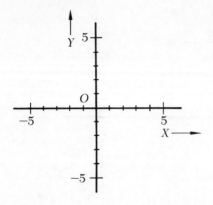

FIGURE 4.8 An X-Y rectangular coordinate system.

Because the grid on which we located the points in these examples is rectangular in shape, these "maps" are called *rectangular coordinate systems.*

We can use this idea derived from position location on maps to construct a generalized rectangular coordinate display that is capable of representing many different types of physical situations. We label the horizontal axis (or *ordinate*) of our rectangular coordinate system with the symbol X, and we label the vertical axis (or *abscissa*) with the symbol Y, as in Figure 4.8. Notice that in Figures 4.6 and 4.7 we specify the location of a point that lies to the left of the origin in terms of the number of miles *west* of O. In Figure 4.8, however, X and Y are allowed to take on *negative* as well as positive values; therefore, a point that lies to the left of O is located in terms of a *negative* value of X.

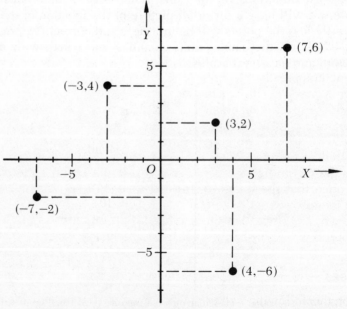

FIGURE 4.9 The location of points (x,y) in the X-Y plane.

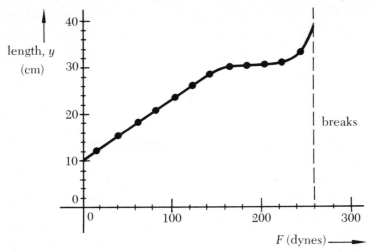

FIGURE 4.10   The length of a rubber band stretched by various forces.

The location of a point in the X-Y coordinate system is specified by stating two numbers, the value of the X coordinate and the value of the Y coordinate; for example ($x = 3$ units, $y = 4$ units). Usually, we simplify this procedure and write only the value of x and the value of y, it being understood that the order of the coordinates is first x and then y. Thus, the point referred to above is written as (3,4); a general point on the plane is written as (x,y). The origin is denoted by (0,0). Figure 4.9 locates several points in this notation.

Notice that we have not specified the magnitude of a unit of measure along the X and Y axes. Each unit could represent 1 cm, or 1 mile, or even 5.8 miles; in fact, the X unit could be different from the Y unit.

The X-Y type of coordinate system is useful for representing various types of physical situations, not only the location of points in terms of the distance from an origin. For example, suppose that we wish to represent graphically the length of a rubber band subject to various stretching forces. In this case, we would let the X-axis represent the *force* (and we would change X to F) and we would let the Y-axis represent the *length*. Figure 4.10 shows the measurements of the length, starting with $y = 10$ cm, the unstretched length, for $F = O$. Values are plotted at increments of 20 dynes of the stretching force and the curve is drawn through the points to give a smooth representation of the data. Notice that the length of the rubber band increases up to about 4 times its unstretched length, at which point it breaks.

We will pursue further the graphical representation of physical data in Chapter 6.

EXERCISES

Plot the following sets of points on X-Y graphs and identify the geometric shapes that the points outline.

1. $(0,0)$, $(2,6)$, $(3,9)$, $(5,15)$ (Ans. 354)

2. $(4,-2)$, $(-1,3)$, $(-1,-2)$, $(4,3)$ (Ans. 137)

3. $(5,-3)$, $(-3,-3)$, $(5,7)$, $(0,-3)$, $(5,2)$ (Ans. 64)

4. $(2,1)$, $(-3,1)$, $(2,7)$, $(-3,7)$ (Ans. 349)

4.4 THREE-DIMENSIONAL COORDINATE SYSTEMS

The coordinate systems discussed in the preceding section are capable of representing the positions of points in a plane (i.e., in *two* dimensions, X and Y). If we wish to locate a point in *space*, such coordinate systems are no longer adequate because we require a third dimension. That is, if X and Y designate horizontal distances, we need also the *height* in order to locate a position in space. For this purpose we must add a third axis, Z, to our coordinate system which now becomes a *three-dimensional* rectangular coordinate system, as shown in Figure 4.11.

There are actually two distinct ways of orienting the X, Y, and Z axes in a three-dimensional coordinate system, as shown in Figure 4.12. That these two orientations are distinct can be seen by the fact that there is no way to turn the system of Figure 4.12*a* in order to make it coincide with the system of Figure 4.12*b*. The only way to convert system (a) into system (b) is to view it in a mirror; that is, system (b) is the *mirror image* of system (a), and vice versa.

The coordinate system of Figure 4.12*a* is called a *right-handed system* because if one imagines that a right-handed screw (an ordinary screw) is turned in the sense that carries the positive X-axis toward the positive Y-axis, then the direction of advance of the screw is in the direc-

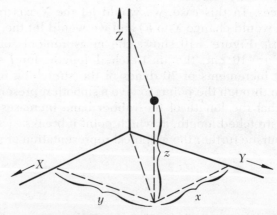

FIGURE 4.11 Location of the point (x,y,z) in a three-dimensional rectangular coordinate system.

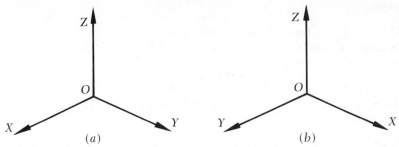

FIGURE 4.12   (a) A *right-handed* coordinate system. (b) A *left-handed* coordinate system.

tion of the positive Z-axis (see Fig. 4.13). The coordinate system of Figure 4.12*b* does not have this property (the direction of advance of a right-handed screw is the direction of the *negative* Z-axis), and such a system is called a *left-handed system*. Right-handed coordinate systems are almost always used for describing physical situations.

The location of a point or the position of a physical object in space requires the specification of *three* numbers—the X, Y, and Z coordinates. If we wish to define a physical *event*, however, we need more information. An *event* is an occurrence that takes place not only at a particular location but also at a particular time. Therefore, the specification of an event requires not only three space coordinates but a *time* coordinate as well. The location of an event (in *space* and in *time*) is given by *four* numbers and the coordinates are (x,y,z,t). That is, our space is actually a *four*-dimensional space with three ordinary space dimensions and one time dimension. Usually we do not think of our space as consisting of four dimensions because there is no apparent coupling or interrelation between ordinary space and time. But in the theory of relativity it becomes clear that there *is* such a coupling and that when objects are in motion with respect to one another with high relative velocities, the coupling of space and time is crucial. We should not lose sight of the fact that our world is, after all, a *four-dimensional* world.

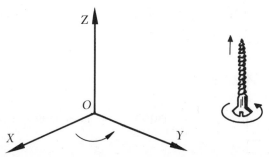

FIGURE 4.13   Definition of a *right-handed* coordinate system in terms of the direction of advance of a right-handed screw.

1. Construct a three-dimensional coordinate system and plot the following points: (3,5,4), (−2,4,−6), (4,−3,7).

2. How would you plot an *event* in a four-dimensional coordinate system? (Ans. 226)

3. Construct a space-time coordinate system for representing *events*. Restrict the space co-ordinates to *two* dimensions (*X* and *Y*). Plot two events that take place at the same location but at different times. Plot two events that take place at the same time but in different locations.

4.5 THE DISTANCE BETWEEN TWO POINTS

One of the important results of plane geometry is the *Pythagorean theorem,* which states that the square of the length of the side opposite the right angle (called the *hypotenuse*) of a right triangle is equal to the squares of the lengths of the other two sides. That is (refer to Figure 4.14),

$$c^2 = a^2 + b^2 \quad \text{(right triangle)} \tag{4.4}$$

Thus, if we know the lengths of any two sides of a right triangle, we can always find the length of the third side. If, in Figure 4.14, we know $c = 14$ cm and $a = 5$ cm, then

$$b = \sqrt{c^2 - a^2} = \sqrt{(14)^2 - (5)^2} \text{ cm} = \sqrt{196 - 25} \text{ cm}$$

$$= \sqrt{171} \text{ cm} = 13.08 \text{ cm}$$

FIGURE 4.14 Identification of sides in a right triangle.

Example 4.5.1

A motorist travels east from a starting point P at a constant speed of 50 mi/hr for 3 hr. He then turns north and for 2 hr he maintains a constant speed of 65 mi/hr. At the end of the 5-hr trip, how far is he from P?

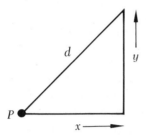

The first leg of the trip is

$$x = vt = (50 \text{ mi/hr}) \cdot (3 \text{ hr}) = 150 \text{ mi}$$

and the second leg is

$$y = (65 \text{ mi/hr}) \cdot (2 \text{ hr}) = 130 \text{ mi}$$

Therefore,

$$d = \sqrt{x^2 + y^2} = \sqrt{(150)^2 + (130)^2} \text{ mi}$$
$$= \sqrt{22500 + 16900} \text{ mi} = \sqrt{39400} \text{ mi}$$
$$= \sqrt{3.94 \cdot 10^2} \text{ mi} = 198.5 \text{ mi}$$

Because the axes of a rectangular coordinate system are at *right* angles, the Pythagorean theorem can be used to calculate the distances between pairs of points in such a system. Consider the point A with coordinates $(3,4)$ shown in Figure 4.15. How far is A from the origin? The point A and the origin define the right triangle OBA. Therefore, the distance c from O to A is

$$c = \sqrt{a^2 + b^2} = \sqrt{(3)^2 + (4)^2} = \sqrt{9 + 16}$$
$$= \sqrt{25} = 5 \text{ units}$$

Notice that this procedure works even if the coordinates of A are *negative* values because it is the *squares* of the coordinates that enter into the calculation.

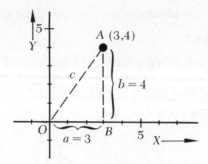

FIGURE 4.15 The distance c from O to A can be found by using the Pythagorean theorem.

If we wish to find the distance between two points neither of which is the origin, then we follow a similar procedure. Such a case is shown in Figure 4.16 in which we require the distance from A to B. The two points A and B and the point C (which has the same x-value as B and the same y-value as A) define the right triangle ACB. Then, $c = \sqrt{a^2 + b^2}$, where

$$a = x_2 - x_1; \quad b = y_2 - y_1 \tag{4.5}$$

Therefore,

$$c = \sqrt{(x_2 - x_1)^2 + (y_2 - y_1)^2} \tag{4.6}$$

In Figure 4.16, the coordinates of the points are

$$A: (x_1, y_1) = (4,2); \quad B: (x_2, y_2) = (8,7)$$

Thus,

$$c = \sqrt{(8 - 4)^2 + (7 - 2)^2} = \sqrt{16 + 25}$$
$$= \sqrt{41} = 6.4 \text{ units}$$

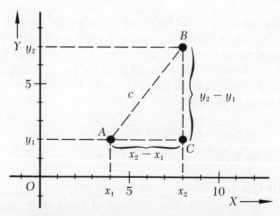

FIGURE 4.16 What is the distance from A to B?

In order to find the distance between two points in three-dimensional space, it is necessary to use the Pythagorean theorem *twice*. This procedure is best illustrated by means of an example.

Example 4.5.2

What is the length of the diagonal of a cube whose sides have a length of 1 m?

The diagonal is the dashed line, OC. First, consider the right triangle OAB and the diagonal OB. Using the Pythagorean theorem, the length OB is

$$OB = \sqrt{(OA)^2 + (AB)^2} = \sqrt{(1)^2 + (1)^2} = \sqrt{2} \text{ m}$$

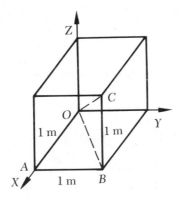

Next, consider the right triangle OBC. Again, we use the Pythagorean theorem to find the length of the side OC (the diagonal of the cube):

$$OC = \sqrt{(OB)^2 + (BC)^2} = \sqrt{(\sqrt{2})^2 + (1)^2} = \sqrt{3} \text{ m}$$

In general, the length of the diagonal of a cube is $\sqrt{3}$ times the length of a side.

The distance between two points located at (x_1, y_1, z_1) and (x_2, y_2, z_2) is found by combining the procedure of the preceding example and the result given in Equation 4.6; we find

$$\text{distance} = \sqrt{(x_2 - x_1)^2 + (y_2 - y_1)^2 + (z_2 - z_1)^2} \tag{4.7}$$

EXERCISES

1. What is the length of the diagonal of a square whose area is 8 cm²?   (Ans. 42)

2. What is the distance between the points $(-3,7)$ and $(-4,-1)$?   (Ans. 160)

3. What is the distance from the origin to the point $(6,4,3)$?   (Ans. 58)

4. What is the length of the diagonal of a cube whose volume is 64 m³?   (Ans. 335)

4.6 RADIAN MEASURE

The most commonly used unit of angular measure is the *degree*, which is 1/360 of a complete circle. For many types of problems in physics and engineering it proves more convenient to use another unit called the *radian* (rad). If we measure the length of arc along the circumference of a circle (see Fig. 4.17), we find that the arc length s is proportional to the angle θ between the two radii that define the arc; that is, $s \propto \theta$. Furthermore, if we hold θ fixed and increase r, then s increases in direct proportion; that is, $s \propto r$. *One radian* is defined to be the angle subtended when the arc length s is exactly equal to the radius r. Thus,

$$s = r\theta \tag{4.8}$$

where θ is measured in radians.

If θ is increased until it is equal to 360°, the arc s is just the circumference, $2\pi r$. Then, $s = 2\pi r = r\theta$, so that $\theta = 2\pi$ radians corresponds to $\theta = 360°$. Therefore,

$$1 \text{ rad} = \frac{360°}{2\pi} = 57\overset{\circ}{.}2958 \cdots \cong 57\overset{\circ}{.}3 \tag{4.9}$$

In order to find the radian equivalent of 1°, we write

$$1° = \frac{2\pi}{360°} = 0.01745 \cdots \text{ rad} \tag{4.10}$$

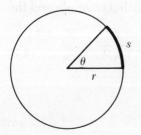

FIGURE 4.17 If $s = r$, the angle θ is equal to 1 radian.

Notice that although the *radian* is a unit of angular measure, it does not have physical dimensions. Therefore, when we put values into Equation 4.8, such as 3 cm = (1 rad) × (3 cm), we do not have different physical dimensions on the two sides of the equation. Usually, we carry the designation *radian* in our equations as a reminder of the angular units we are using, but when the final answer is obtained, it is sufficient to include only the *physical* dimensions. For example, if we calculate the length of the arc of a circle with $r = 10$ cm that is intercepted by an angle $\theta = 0.5$ radian, we obtain

$$s = r\theta = (10 \text{ cm}) \times (0.5 \text{ rad}) = 5 \text{ cm}$$

Example 4.6.1

A 20-cm length of string is wrapped around a pipe and it is found that the ends of the string intercept an angle of 72° at the center of the pipe. What is the radius of the pipe?
First, the angle is

$$\theta = 72° = \frac{72°}{57°3} \text{ rad} = 1.26 \text{ rad}$$

Then,

$$r = \frac{s}{\theta} = \frac{20 \text{ cm}}{1.26 \text{ rad}} = 15.9 \text{ cm}$$

The fact that $s = r\theta$ provides a method for closely estimating distances in certain circumstances. Suppose that a pole is placed vertically in the ground a distance R away from an observer located at a point O (Fig. 4.18). The observer measures the angle subtended by the pole and finds it to be θ. Mentally, the observer constructs a circle of radius R centered at O and passing through the bottom of the pole. The arc length S is

$$S = R\theta \qquad\qquad (4.11)$$

Since θ is a small angle, S is approximately equal to the height of the pole h, and we can write

$$h \cong R\theta \qquad\qquad (4.12)$$

FIGURE 4.18 Estimating the height of an object by measuring the angle subtended at O; $h \cong R\theta$.

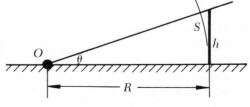

This method for obtaining approximate values for h will be useful if θ is sufficiently small. Even for θ as large as 20°, the error is only about 4 per cent. If θ is a few degrees, the error is usually negligible for most purposes; for $\theta = 1°$, the error is 0.01 per cent or 1 part in 10^4. (Remember that when using Equation 4.11 or 4.12 the angle θ must be expressed in *radians*.)

For a further discussion of this *small-angle approximation,* see Section 6.5.

Example 4.6.2

An observer on the Mall looks at the Washington Monument (height = 555 ft) and determines that the Monument subtends an angle of 12°. How far away (approximately) is the observer from the Monument?

The angle is

$$\theta = 12° = \frac{12°}{57°3} \text{ rad} = 0.21 \text{ rad}$$

Then,

$$R \cong \frac{h}{\theta} = \frac{555 \text{ ft}}{0.21 \text{ rad}} = 2.64 \times 10^3 \text{ ft}$$

or about 1/2 mile. The error in this approximate result is less than 1 per cent.

EXERCISES

1. $180° = \underline{\hspace{1cm}}$ rad   (Ans. 51)

2. $45° = \underline{\hspace{1cm}}$ rad   (Ans. 47)

3. $3 \text{ rad} = \underline{\hspace{1cm}}°$ (Ans. 320)

4. What angle does a meter stick subtend at a distance of 100 m?   (Ans. 330)

5. The Moon subtends an angle of approximately $\frac{1°}{2}$ at the Earth and the Earth-Moon distance is approximately 240,000 mi. What is the diameter of the Moon?   (Ans. 228)

6. The Empire State Building is 1245 ft high. From a certain point on Long Island an observer finds that the top of the building is 3° above the horizon. How far is the observer from midtown Manhattan?   (Ans. 125)

4.7 ANGULAR MOTION

If an object moves in a circular path with a speed such that one complete revolution requires 1 second, we say that the object moves at an angular rate of 1 rev/s. Since one complete revolution corresponds to 2π radians, we can alternatively say that the object moves with an *angular velocity* of 2π rad/s. It is customary to denote angular velocity (measured in rad/s) by the symbol ω (omega).

If we do not have a complete revolution on which to base a calculation we can still define the angular velocity in a manner entirely analogous to that used for ordinary (or *linear*) velocity. Thus, if an object moves uniformly from a point identified by the angle θ_1 to a point identified by θ_2 in a time interval $t_2 - t_1$ (see Fig. 4.19), the *angular velocity* is

$$\omega = \frac{\theta_2 - \theta_1}{t_2 - t_1} = \frac{\Delta\theta}{\Delta t} \tag{4.13}$$

Since it is only the angular *difference* that is important, the position labeled $\theta = 0$ is arbitrary.

The *period* of circular motion is the time required for one complete revolution or cycle of the motion. The period and the angular velocity are inversely related, since the greater the angular velocity, the shorter the time required to make a revolution. The period is denoted by the symbol τ (tau):

$$\tau = \frac{2\pi}{\omega} \tag{4.14}$$

If an object moves with uniform speed in a circular path with radius r, the distance traveled in 1 period is just the circumference of the circle, $2\pi r$. The time required for this motion is τ. Therefore, the magnitude of the velocity is

$$v = \frac{\text{distance}}{\text{time}} = \frac{2\pi r}{2\pi/\omega}$$

Thus,

$$v = r\omega \tag{4.15}$$

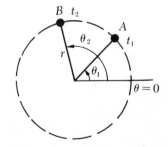

FIGURE 4.19 If the object moves uniformly from A to B, the angular velocity is $(\theta_2 - \theta_1)/(t_2 - t_1)$ rad/s.

Example 4.7.1

An automobile moves with a constant speed of 50 mi/hr around a circular track which has a diameter of 1 mi. What is the angular velocity and the period of the motion?

$$\omega = \frac{v}{r} = \frac{50 \text{ mi/hr}}{0.5 \text{ mi}} = 100 \text{ rad/hr}$$

$$\tau = \frac{2\pi}{\omega} = \frac{2\pi}{100 \text{ rad/hr}} = 0.063 \text{ hr} = 3.8 \text{ min}$$

EXERCISES

1. An object moves in a circular path and completes 40 revolutions in 2.8 s. What is the angular velocity?   (Ans. 294)

2. The shaft of a certain electric motor rotates at a rate of 1400 rpm. What is the angular velocity in rad/s?   (Ans. 278)

3. In Exercise 2, what is the period of rotation of the shaft?   (Ans. 157)

4. An object moves in a circular path with a radius of 40 cm and after a time of 2 hr the object has moved through a total distance of 1 mi. What is the angular velocity of the object?   (Ans. 258)

5. What is the angular velocity of the Earth's rotation around its axis?   (Ans. 53)

6. What is the speed of a point on the surface of the Earth at the equator due to the rotation of the Earth around its axis? (Use the fact that the radius of the Earth is approximately 4000 mi and refer to the result of Exercise 5). Express the result in mi/s.   (Ans. 298)

CHAPTER 5

TRIGONOMETRY

5.1 SINES, COSINES, AND TANGENTS

A *right-triangle* is a triangle in which one of the angles is 90° (or $\pi/2$ radians). Such a right triangle is shown in Figure 5.1 in which the angle $\angle ACB$ is equal to 90°.

The lengths of the sides opposite the vertices A, B, and C are, respectively, a, b, and c. From the Pythagorean theorem (see Section 4.5), a, b, and c are related by

$$c^2 = a^2 + b^2 \tag{5.1}$$

We denote by θ the angle $\angle BAC$, and by ϕ the angle $\angle ABC$. The ratio of the length of the side opposite θ to the length of the hypotenuse (that is, side AB) is called the *sine* of the angle θ. This quantity is abbreviated as sin θ. Therefore,

$$\sin \theta = \frac{\text{side opposite angle } \theta}{\text{hypotenuse}} = \frac{a}{c} \tag{5.2}$$

FIGURE 5.1   Right triangle with $\angle ACB = 90°$.

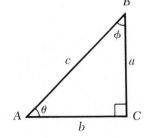

Similarly, the ratio b/c is defined to be the *cosine* of θ; that is,

$$\cos \theta = \frac{\text{side adjacent angle } \theta}{\text{hypotenuse}} = \frac{b}{c} \qquad (5.3)$$

Using Equation 5.1 we can also express $\sin \theta$ and $\cos \theta$ exclusively in terms of a and b:

$$\sin \theta = \frac{a}{\sqrt{a^2 + b^2}}; \quad \cos \theta = \frac{b}{\sqrt{a^2 + b^2}} \qquad (5.4)$$

The ratio of the length of the side opposite θ to the length of the adjacent side is called the *tangent* of the angle θ:

$$\tan \theta = \frac{\text{side opposite angle } \theta}{\text{side adjacent angle } \theta} = \frac{a}{b} \qquad (5.5)$$

The tangent of the angle θ is not independent of the sine and cosine since

$$\tan \theta = \frac{a}{b} = \frac{a/c}{b/c} = \frac{\sin \theta}{\cos \theta}$$

That is,

$$\tan \theta = \frac{\sin \theta}{\cos \theta} \qquad (5.6)$$

In general, since b and a are each less than $c = \sqrt{a^2 + b^2}$, it follows from Equation 5.4 that $\sin \theta$ and $\cos \theta$ *do not exceed unity*. However, since $\tan \theta = b/a$, the tangent of the angle θ will exceed unity whenever b is larger than a.

The quantities $\sin \theta$, $\cos \theta$, and $\tan \theta$ are all called *trigonometric functions* of the angle θ.

Example 5.1.1.

Consider a right triangle with $a = 4$ cm and $b = 3$ cm. In this case,

$$c = \sqrt{4^2 + 3^2} \text{ cm} = \sqrt{25} \text{ cm} = 5 \text{ cm}$$

and the triangle has the form shown in the accompanying figure. Therefore, from the defining equations for $\sin \theta$, $\cos \theta$, and $\tan \theta$, we find

$$\sin \theta = \frac{4 \text{ cm}}{5 \text{ cm}} = 0.800, \quad \cos \theta = \frac{3 \text{ cm}}{5 \text{ cm}} = 0.600, \quad \tan \theta = \frac{4 \text{ cm}}{3 \text{ cm}} = 1.333$$

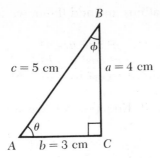

Note that the units of length cancel in the preceding expressions so that the resulting numbers for sin θ, cos θ, and tan θ are *dimensionless*. Note also that tan θ exceeds unity for this example.

EXERCISES

If $a = b = 1$ ft, then

1. $c =$ ____   (Ans. 193)

2. $\sin \theta =$ ____   (Ans. 331)

3. $\cos \theta =$ ____   (Ans. 324)

4. $\tan \theta =$ ____   (Ans. 166)

If $b = 6$ m and $c = 10$ m, then

5. $a =$ ____   (Ans. 318)

6. $\sin \theta =$ ____   (Ans. 78)

7. $\cos \theta =$ ____   (Ans. 108)

8. $\tan \theta =$ ____   (Ans. 203)

(Hint: Since b and c are known, a can be determined by writing the Pythagorean theorem in the form $a^2 = c^2 - b^2$.)

5.2 TRIGONOMETRIC IDENTITIES

We now verify some important relationships between the sines and cosines of angles in right triangles. Referring to Figure 5.1 and the defining equations for sin θ and cos θ, we note that

$$c \sin \theta = a$$

$$c \cos \theta = b$$

If we square these equations and add them, we find[*]

$$c^2 \sin^2 \theta = a^2$$
$$\frac{c^2 \cos^2 \theta = b^2}{c^2(\sin^2 \theta + \cos^2 \theta) = a^2 + b^2}$$

Comparing this result with Equation 5.1, we see that

$$\sin^2 \theta + \cos^2 \theta = 1 \qquad (5.7)$$

Therefore, if we know the value of sin θ, then cos θ can be determined from Equation 5.7, and *vice versa*. For example, if

$$\sin \theta = \frac{4}{5}$$

then it follows from Equation 5.7 that

$$\cos^2 \theta = 1 - \left(\frac{4}{5}\right)^2 = 1 - \frac{16}{25} = \frac{9}{25}$$

Taking the positive square root we find

$$\cos \theta = \frac{3}{5}$$

which is in agreement with the result for the example discussed in Section 5.1.

There are some useful trigonometric relationships between the angles θ and ϕ in the right triangle shown in Figure 5.1. First, since the angles in a triangle add to 180°, we note that

$$\phi + \theta + 90° = 180°.$$

Therefore, the angles ϕ and θ are related by

$$\phi = 90° - \theta \qquad (5.8)$$

The angle ϕ is called the *complement* of the angle θ. Similarly, since $\theta = 90° - \phi$, the angle θ is the complement of ϕ. The sine, cosine, and tangent of the angle ϕ are, by definition,

$$\sin \phi = \frac{b}{c}; \quad \cos \phi = \frac{a}{c}; \quad \tan \phi = \frac{b}{a}$$

---

[*]It is customary to write $(\sin \theta)^2 = \sin^2 \theta$, and $(\cos \theta)^2 = \cos^2 \theta$. This is just a notational device, and no extra meaning should be read into it.

Comparing these equations with the equations for sin θ, cos θ, and tan θ, we note that

$$\sin \theta = \frac{a}{c} = \cos \phi$$

$$\cos \theta = \frac{b}{c} = \sin \phi \qquad (5.9)$$

$$\tan \theta = \frac{a}{b} = \frac{1}{\tan \phi}$$

Combining Equations 5.8 and 5.9 we find the identities,

$$\sin \theta = \cos (90° - \theta)$$

$$\cos \theta = \sin (90° - \theta) \qquad (5.10)$$

$$\tan \theta = \frac{1}{\tan (90° - \theta)}$$

That is, *the sine of an angle is equal to the cosine of the complement of that angle; the cosine of an angle is equal to the sine of the complement of that angle; and so on.* Thus, for example,

$$\sin 20° = \cos 70°$$

$$\cos 50° = \sin 40°$$

$$\tan 80° = \frac{1}{\tan 10°}$$

$$\cos 63.5° = \sin 26.5°$$

EXERCISES

1. If sin $\theta = 2/3$, then cos^2 $\theta =$ ____   (Ans. 260)

2. If sin $\theta = 2/3$, then cos $\theta =$ ____   (Ans. 171)

3. If cos $\theta = 0.5$, then sin^2 $\theta =$ ____   (Ans. 43)

4. If cos $\theta = 1/\sqrt{2}$, then sin $\theta =$ ____   (Ans. 302)

5. The complement of 74° = ____   (Ans. 44)

6. The complement of $\pi/4$ radians = ____   (Ans. 283)

7. sin 15° = cos ____   (Ans. 169)

8. cos 75° = sin ____   (Ans. 95)

9. sin 45° = cos ____   (Ans. 287)

10. tan 81° = $\dfrac{1}{\tan \text{____}}$ (Ans. 168)

5.3 IMPORTANT TRIANGLES AND THE USE OF TRIGONOMETRIC TABLES

If the lengths of the sides of the right triangle in Figure 5.1 are known, then the evaluation of sin θ, cos θ, and tan θ (or sin ϕ, cos ϕ, and tan ϕ) proceeds in the manner described in Sections 5.1 and 5.2. In this section we discuss methods for

(a) evaluating sin θ, cos θ, and tan θ for circumstances in which *the value of θ is specified* (in degrees or radians), and

(b) evaluating θ for circumstances in which sin θ, cos θ, or tan θ is specified.

We first consider the special right-angle triangles shown in Figures 5.2 and 5.3. As indicated in Figure 5.2, the lengths of the sides of the 45°–45°–90° triangle stand in the ratio

$$1 : 1 : \sqrt{2}$$

For the 30°–60°–90° triangle the lengths of the sides stand in the ratio

$$1 : \sqrt{3} : 2$$

as shown in Figure 5.3. Thus, it is apparent from the defining equations for the sine, cosine, and tangent functions that

$$\sin 45° = \frac{1}{\sqrt{2}} = 0.707$$

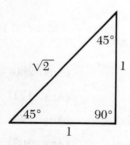

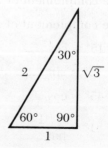

FIGURE 5.2   The 45°–45°–90° triangle.

FIGURE 5.3   The 30°–60°–90° triangle.

$$\cos 45° = \frac{1}{\sqrt{2}} = 0.707$$

$$\tan 45° = \frac{1}{1} = 1$$

and

$$\sin 30° = \cos 60° = \frac{1}{2} = 0.5$$

$$\cos 30° = \sin 60° = \frac{\sqrt{3}}{2} = 0.866$$

$$\tan 30° = \frac{1}{\sqrt{3}} = 0.577$$

$$\tan 60° = \frac{\sqrt{3}}{1} = 1.732$$

Table 5.1 summarizes the values of the trigonometric functions for all the angles involved in these simple triangles, plus the cases $\theta = 0°$ and $\theta = 90°$. The symbol ∞ in Table 5.1 means *indefinitely large* or *infinite*. Any non-zero number divided by 0 is infinite. Referring to Figure 5.1, we note that for $\theta = 90°$ we require $b = 0$. Therefore, $\tan 90° = a/0 = \infty$, as listed in Table 5.1.

Of course, Table 5.1 is by no means complete. For example, it is of no assistance in calculating $\cos 21°$. Given in Table I (pages 193–210) is a much more extensive table of values for $\sin \theta$, $\cos \theta$, and $\tan \theta$, for all angles ranging from $\theta = 0°$ to $\theta = 90°$ in steps of $0°.1$. In order to de-

TABLE 5.1   VALUES OF SOME TRIGONOMETRIC FUNCTIONS

	$\sin \theta$	$\cos \theta$	$\tan \theta$
$\theta = 0°$	0	1	0
$\theta = 30°$	0.5	$\frac{\sqrt{3}}{2} = 0.866$	$\frac{1}{\sqrt{3}} = 0.577$
$\theta = 45°$	$\frac{1}{\sqrt{2}} = 0.707$	$\frac{1}{\sqrt{2}} = 0.707$	1
$\theta = 60°$	$\frac{\sqrt{3}}{2} = 0.866$	0.5	$\frac{\sqrt{3}}{1} = 1.732$
$\theta = 90°$	1	0	∞

termine sin 10°, for example, we read down the left-hand column on page 195 to $\theta = 10°$, and then read across to the sine column to find*

$$\sin 10° = 0.174$$

In a similar manner it is easy to verify that

$$\sin 83°\!.5 = 0.994, \quad \cos 21° = 0.934, \quad \tan 15°\!.5 = 0.277$$

and so on. (Look up these values in Table I to make certain that you understand the use of the table.)

For problems in which the value of sin θ, cos θ, or tan θ is specified, the trigonometric tables can also be used (in the reverse fashion to that described above) to evaluate the angle θ. For example, if sin $\theta = 0.707$, then, from Table 5.1, $\theta = 45°$. Similarly, it follows from Table I, pages 198, 202, and 204, that

$$\sin \theta = 0.438 \text{ corresponds to } \theta = 26°$$

$$\cos \theta = 0.669 \text{ corresponds to } \theta = 48°$$

$$\tan \theta = 1.664 \text{ corresponds to } \theta = 59°$$

(Again, look up these values in the table.)

---

*In Table I, pages 193–210, the values of sin θ, cos θ, and tan θ are tabulated to *four* decimal places (except that tan θ is given to five significant figures for θ near 90°). However, throughout the worked examples the values of trigonometric functions are, for the most part, expressed to *three* decimal places. Thus, for example, in evaluating sin 10° from Table I, page 195, we round off sin 10° = 0.1736 to three decimal places, which gives sin 10° = 0.174.

Example 5.3.1

Consider the right triangle shown in the figure below. This is a particularly simple triangle because the sides stand in the ratio 3 : 4 : 5. (Notice that $5^2 = 4^2 + 3^2$.) Evidently,

$$\cos \theta = \frac{3}{5} = 0.600$$

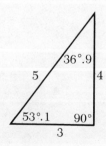

From Table I, page 203, we conclude that

$$\theta = 53°1$$

Since ϕ is the complement of θ, $\phi = 90° - \theta = 36°9$. Thus, as shown in the figure, the angles in a right triangle whose sides stand in the ratio 3 : 4 : 5 are 36°9, 53°1, and 90°.*

---

*The values, 36°9 and 53°1, are *not* exact. To five-decimal accuracy these angles are 36°86848 and 53°13152, respectively.

Example 5.3.2

A surveyor wishes to determine the distance between two points, A and B, but he cannot make a direct measurement because a river intervenes. How can he obtain a precise value for the distance?

The surveryor first selects a point C and he adjusts the location of C until a sighting with his transit (located at A) shows that the lines AB and AC are at right angles; that is, $\angle CAB = 90°$. He then measures the distance from A to C and finds

$$AC = 264 \text{ ft}$$

Next, he positions his transit at C and sights toward A and then toward B, measuring the angle between the lines CA and CB. He finds $\angle ACB = 62°3$

Now,

$$\tan \angle ACB = \frac{AB}{AC}$$

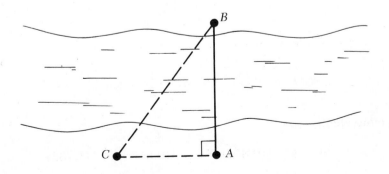

or, solving for the desired distance AB,

$$AB = AC \tan \angle ACB = (264 \text{ ft}) \times (\tan 62°3)$$
$$= (264 \text{ ft}) \times (1.905) = 503 \text{ ft}$$

Notice that the actual distance from A to C does not matter. That is, as long as $\angle CAB = 90°$ and the points A and B are fixed, the location of point C is irrelevant. The surveyor needs only to measure the distance AC and the angle $\angle ACB$; any combination will then produce the same result for AB.

The operation of expressing an angle θ in terms of the value of its sine, cosine, or tangent is stated mathematically in the following manner. Referring to the right triangle in Figure 5.1, we can express θ as

$$\theta = \sin^{-1}\left(\frac{a}{c}\right) \tag{5.11}$$

This equation is equivalent to the statement:

$$\theta = \text{the angle whose sine is } \frac{a}{c} \tag{5.12}$$

The exponent -1 on the sine function in Equation 5.11 does *not* refer to the reciprocal; it is nothing more than a notational device that expresses the statement in Equation 5.12.

Since $\tan\theta = a/b$, and $\cos\theta = b/c$, we can also express θ as

$$\theta = \tan^{-1}\left(\frac{a}{b}\right) \tag{5.13}$$

or as

$$\theta = \cos^{-1}\left(\frac{b}{c}\right) \tag{5.14}$$

As an example, we note from Figure 5.3 (or Table 5.1) that

$$60° = \sin^{-1}\left(\frac{\sqrt{3}}{2}\right) = \cos^{-1}\left(\frac{1}{2}\right) = \tan^{-1}\left(\frac{\sqrt{3}}{1}\right)$$

or, equivalently,

$$60° = \sin^{-1}(0.866) = \cos^{-1}(0.5) = \tan^{-1}(1.732)$$

Exercises

1. $\sin 1° =$ ____   (Ans. 120)

2. $\cos 89° =$ ____   (Ans. 30)

3. $\tan 1° =$ ____   (Ans. 112)

4. $\tan 55° =$ ____   (Ans. 62)

5. $\cos 73°2 =$ ____   (Ans. 37)

6. $1.33 = \tan$ ____   (Ans. 133)

7. $0.575 = \sin$ ____   (Ans. 60)

8. $0.600 = \cos$ ____   (Ans. 10)

9. $\sin^{-1}(0.5) =$ ____   (Ans. 284)

10. $\cos^{-1}(0.665) =$ ____   (Ans. 246)

11. $\tan^{-1}(1) =$ ____   (Ans. 11)

5.4 POLAR COORDINATES

In Figure 5.4 the point P is located in Quadrant I of a rectangular coordinate system. As discussed in Section 4.2, the location of P can be specified by a pair of numbers (x,y), where x is the projection of OP along the horizontal X-axis, and y is the projection of OP along the vertical Y-axis.

In many cases it is more convenient to specify the location of P in terms of *polar coordinates* (r,θ) instead of rectangular coordinates (x,y).

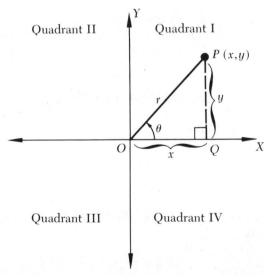

FIGURE 5.4 Polar coordinates (r,θ) for P in Quadrant I.

As indicated in Figure 5.4, the *radial coordinate* r is defined to be the distance from the origin O to the point P. Since OP forms the hypotenuse of the right triangle OPQ, it follows from the Pythagorean theorem that

$$r = \sqrt{x^2 + y^2} \qquad (5.15)$$

Furthermore, the *angular coordinate* θ is defined to be the angle between the positive X-axis and the line OP, measured in a counter-clockwise sense. Referring to Figure 5.4 and the defining equations for $\sin \theta$ and $\cos \theta$ we find that x, r, and θ are related by

$$x = r \cos \theta \qquad (5.16)$$

and y, r, and θ are related by

$$y = r \sin \theta \qquad (5.17)$$

Equations 5.15 to 5.17 relate the rectangular coordinates (x, y) to the polar coordinates (r, θ). Note that Equations 5.16 and 5.17 can be combined to give

$$\frac{y}{x} = \frac{r \sin \theta}{r \cos \theta} = \tan \theta$$

That is,

$$\tan \theta = \frac{y}{x} \qquad (5.18)$$

which directly relates θ to the coordinates x and y.

If values for x and y are given, then Equations 5.15 and 5.18 can be used to determine r and θ, respectively. For example, if $x = 3$ cm and $y = 4$ cm, then

$$r = \sqrt{3^2 + 4^2} \text{ cm} = 5 \text{ cm}$$

Furthermore, from Equation 5.18,

$$\tan \theta = \frac{4 \text{ cm}}{3 \text{ cm}} = 1.33$$

Comparing with the figure in Example 5.3.1 or making use of Table I, page 203, we find

$$\theta = 53^\circ.1$$

Example 5.4.1

A particle is located at the point ($x = 10$ cm, $y = 20$ cm) in a rectangular coordinate system. What are the corresponding polar coordinates?

$$r = \sqrt{(10)^2 + (20)^2} \text{ cm} = \sqrt{500} \text{ cm} = 22.36 \text{ cm}$$

We also have

$$\tan \theta = \frac{y}{x} = \frac{20 \text{ cm}}{10 \text{ cm}} = 2$$

Therefore,

$$\theta = \tan^{-1} 2 = 63°4$$

where we have used the Trigonometric Tables and found tan $63°4 = 1.997$.

Example 5.4.2

The polar coordinates of an object are ($r = 10$ ft, $\theta = 60°$). What are the corresponding rectangular coordinates?

$$x = r \cos \theta = (10 \text{ ft}) \cos 60° = (10 \text{ ft}) \times 0.5 = 5 \text{ ft}$$

$$y = r \sin \theta = (10 \text{ ft}) \sin 60° = (10 \text{ ft}) \times 0.866 = 8.66 \text{ ft}$$

We now extend our discussion of polar coordinates to situations in which P is located in Quadrants II, III, or IV (see Figures 5.5 to 5.7).

Note that the angular coordinate θ is always measured in a *counterclockwise sense from the positive X-axis*, no matter in which quadrant P is located. Furthermore, as in Quadrant I, r is equal to the distance between the origin O and the point P. In Quadrants II to IV, the rectangular coordinates (x,y) and the polar coordinates (r,θ) are related in the same manner as in Quadrant I. That is,

$$x = r \cos \theta$$
$$y = r \sin \theta \qquad\qquad (5.19)$$
$$\tan \theta = \frac{y}{x}$$

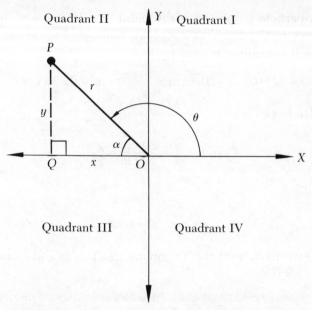

FIGURE 5.5   Polar coordinates (r, θ) for P in Quadrant II.

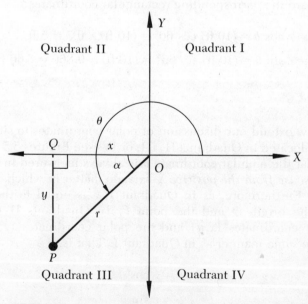

FIGURE 5.6   Polar coordinates (r, θ) for P in Quadrant III.

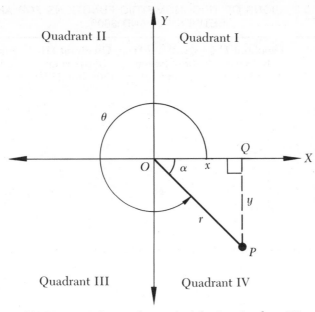

FIGURE 5.7 Polar coordinates (r,θ) for P in Quadrant IV.

In Equations 5.19 and Figures 5.4 to 5.7, $r = \sqrt{x^2 + y^2}$ is a positive number. Table 5.2 summarizes the signs of x and y in the various quadrants.

Taking into account the signs of x and y in Equations 5.19, it is apparent that sin θ, cos θ, and tan θ may be positive or negative, depending on the quadrant. For example, in Quadrant II, x is negative and y is positive. Therefore, since r is positive, we conclude from Equations 5.19 that

$$\left.\begin{array}{l}\cos \theta \text{ is negative} \\ \sin \theta \text{ is positive} \\ \tan \theta \text{ is negative}\end{array}\right\} \text{ in Quadrant II}$$

The signs of sin θ, cos θ, and tan θ can be determined in an analogous

TABLE 5.2 SIGNS OF x AND y FOR QUADRANTS I TO IV

	Quadrant I (θ between 0° and 90°)	Quadrant II (θ between 90° and 180°)	Quadrant III (θ between 180° and 270°)	Quadrant IV (θ between 270° and 360°)
Sign of x	+	−	−	+
Sign of y	+	+	−	−

TABLE 5.3 SIGNS OF TRIGONOMETRIC FUNCTIONS FOR ANGLES BETWEEN 0° AND 360°

	Quadrant I (θ between 0° and 90°)	Quadrant II (θ between 90° and 180°)	Quadrant III (θ between 180° and 270°)	Quadrant IV (θ between 270° and 360°)
$\sin \theta = \dfrac{y}{r}$	+	+	−	−
$\cos \theta = \dfrac{x}{r}$	+	−	−	+
$\tan \theta = \dfrac{y}{x}$	+	−	+	−

manner for the remaining quadrants. The results are summarized in Table 5.3.

EXERCISES

1. For $x = 2$ cm and $y = 5$ cm, $r = $ ____   (Ans. 152)

2. For $x = 3$ cm and $y = 4$ cm, $\sin \theta = $ ____   (Ans. 329)

3. For $x = 5$ m and $y = 5$ m, $\theta = $ ____   (Ans. 23)

4. For $r = 10$ in and $x = 5$ in, $y = $ ____   (Ans. 192)

5. For $y = 50$ cm and $x = 100$ cm, $\tan \theta = $ ____   (Ans. 45)

6. For $x = -6$ cm and $y = 8$ cm, $\tan \theta = $ ____   (Ans. 207)

7. For $x = -4$ km and $r = 11$ km, $\cos \theta = $ ____   (Ans. 317)

8. For $r = 10$ m and $\theta = 60°$, $x = $ ____   (Ans. 319)

9. For $r = 80$ m and $\theta = 15°$, $y = $ ____   (Ans. 241)

10. For $r = 60$ mi and $\theta = 80°$, $x = $ ____   (Ans. 264)

11. The sign of $\sin 10°$ is ____   (Ans. 170)

12. The sign of $\tan 190°$ is ____   (Ans. 358)

13. The sign of $\cos 300°$ is ____   (Ans. 210)

14. The sign of $\sin 102°$ is ____   (Ans. 73)

15. The sign of $\tan 359°$ is ____   (Ans. 190)

5.5 TRIGONOMETRIC FUNCTIONS FOR ANGLES GREATER THAN 90°

When the angle θ exceeds 90°, the values of $\sin \theta$, $\cos \theta$, and $\tan \theta$ can always be expressed in terms of $\sin \alpha$, $\cos \alpha$, and $\tan \alpha$, where α is the angle $\angle QOP$ shown in Figures 5.5 to 5.7. Note that α is always an *acute angle*; that is, α is *less than* 90°. We consider each quadrant separately:

Quadrant II — Angles θ between 90° and 180°

When θ is between 90° and 180°, we note from Figure 5.5 that $\theta + \alpha = 180°$, or equivalently,

$$\alpha = 180° - \theta \qquad (5.20)$$

The angle $\alpha = 180° - \theta$ is called the *supplement* of the angle θ. From Equation 5.19, sin θ is equal to y/r. Referring to triangle OPQ in Figure 5.5, we see that sin α is also equal to y/r. Therefore, we conclude that

$$\sin \theta = \sin \alpha \qquad (5.21)$$

To evaluate cos θ in Quadrant II requires some care. From Table 5.3 we note that cos $\theta = x/r$ is negative in Quadrant II. Since $-x$ is positive in Quadrant II, it is evident from Figure 5.5 that cos $\alpha = -x/r$ is positive. Therefore, we conclude that

$$\cos \theta = -\cos \alpha \qquad (5.22)$$

That is, the cosine of an angle is equal to the *negative* of the cosine of its supplement. Combining Equations 5.21 and 5.22, we also find

$$\frac{\sin \theta}{\cos \theta} = \frac{\sin \alpha}{-\cos \alpha}$$

or,

$$\tan \theta = -\tan \alpha \qquad (5.23)$$

Substituting $\alpha = 180° - \theta$ into Equations 5.21 to 5.23, the results can be summarized as

For θ between 90° and 180° (Quadrant II):

$$\sin \theta = \sin (180° - \theta)$$
$$\cos \theta = -\cos (180° - \theta) \qquad (5.24)$$
$$\tan \theta = -\tan (180° - \theta)$$

Keep in mind that the angle $\alpha = 180° - \theta$ is less than 90° when θ is between 90° and 180°. Equations 5.24 are therefore very useful, since sin θ, cos θ, and tan θ are expressed in terms of trigonometric functions of angles less than 90°, which are tabulated in Table I on pages 193–210.

As an example, for $\theta = 115°$ we find, from the first equation in Equations 5.24,

$$\sin 115° = \sin(180° - 115°) = \sin 75°$$

From Table I, page 208, we obtain sin 75° = 0.966, which gives

$$\sin 115° = 0.966$$

Similarly, from Equations 5.24,

$$\cos 115° = -\cos(180° - 115°) = -\cos 75° = -0.259$$
$$\tan 115° = -\tan(180° - 115°) = -\tan 75° = -3.732$$

As a further example, for $\theta = 150°$ we find

$$\sin 150° = \sin(180° - 150°) = \sin 30° = 0.500$$
$$\cos 150° = -\cos(180° - 150°) = -\cos 30° = -0.866$$
$$\tan 150° = -\tan(180° - 150°) = -\tan 30° = -0.577$$

Quadrant III—Angles θ between 180° and 270°

It is evident from Figure 5.6 that in Quadrant III the angles α and θ are related by $\theta = 180° + \alpha$, that is,

$$\alpha = \theta - 180° \tag{5.25}$$

Making use of Equation 5.19, the sign relations in Table 5.3, and the defining equations for sin α, cos α, and tan α, we find

$$\sin \theta = -\sin \alpha$$
$$\cos \theta = -\cos \alpha$$
$$\tan \theta = \tan \alpha$$

Since $\alpha = \theta - 180°$, these results can be summarized as

For θ between 180° and 270° (Quadrant III):

$$\sin \theta = -\sin(\theta - 180°)$$
$$\cos \theta = -\cos(\theta - 180°) \tag{5.26}$$
$$\tan \theta = \tan(\theta - 180°)$$

When θ is between 180° and 270°, the angle $\alpha = \theta - 180°$ is less than 90°. Therefore, Equation 5.26 expresses sin θ, cos θ, and tan θ in terms of trigonometric functions of angles less than 90°.

For example, to calculate sin 250°, the first equation of Equations 5.26 yields

$$\sin 250° = -\sin(250° - 180°) = -\sin 70°$$

From Table I, page 207, we find sin 70° = 0.940. Therefore,

$$\sin 250° = -0.940$$

Similarly, from Equations 5.26,

$$\cos 250° = -\cos(250° - 180°) = -\cos 70° = -0.342$$
$$\tan 250° = \tan(250° - 180°) = \tan 70° = 2.748$$

Quadrant IV — Angles θ between 270° and 360°

In Quadrant IV it is apparent from Figure 5.7 that the angles α and θ are related by $\theta + \alpha = 360°$, that is,

$$\alpha = 360° - \theta \qquad (5.27)$$

Making use of Equation 5.19, the sign relations in Table 5.3, and the defining equations for sin α, cos α, and tan α in Quadrant IV, we find

$$\sin \theta = -\sin \alpha$$
$$\cos \theta = \cos \alpha$$
$$\tan \theta = -\tan \alpha$$

Since $\alpha = 360° - \theta$, these results can be summarized as

For θ between 270° and 360° (Quadrant IV):

$$\sin \theta = -\sin(360° - \theta)$$
$$\cos \theta = \cos(360° - \theta) \qquad (5.28)$$
$$\tan \theta = -\tan(360° - \theta)$$

When θ is between 270° and 360°, the angle $\alpha = 360° - \theta$ is less than 90°. Therefore, Equation 5.28 expresses sin θ, cos θ, and tan θ in terms of trigonometric functions of angles less than 90°.

As an example, for $\theta = 315°$, we find, from Equations 5.28,

$$\sin 315° = -\sin(360° - 315°) = -\sin 45° = -0.707$$
$$\cos 315° = \cos(360° - 315°) = \cos 45° = 0.707$$
$$\tan 315° = -\tan(360° - 315°) = -\tan 45° = -1.000$$

As a further example, for $\theta = 359°$ we find, from Equations 5.28,

$$\sin 359° = -\sin(360° - 359°) = -\sin 1° = -0.0175$$
$$\cos 359° = \cos(360° - 359°) = \cos 1° = 0.9998$$
$$\tan 359° = -\tan(360° - 359°) = -\tan 1° = -0.0175$$

EXERCISES

1. $\tan 108° =$ _____   (Ans. 111)

2. $\cos 179° =$ _____   (Ans. 251)

3. $\sin 160°5 =$ _____   (Ans. 311)

4. $\sin 255° =$ _____   (Ans. 65)

5. $\cos 243° =$ _____   (Ans. 216)

6. $\tan 239°5 =$ _____   (Ans. 184)

7. $\tan 281° =$ _____   (Ans. 100)

8. $\cos 300°3 =$ _____   (Ans. 136)

9. $\sin 359° =$ _____   (Ans.   3)

5.6   SMALL ANGLE APPROXIMATIONS

In this section we derive approximate expressions for $\sin \theta$, $\cos \theta$, and $\tan \theta$, for situations in which the angle θ is much smaller than 90°. Consider the right triangle shown in Figure 5.8. We assume that side BC is much shorter in length than side AC; that is, $y \ll x$. In this case the angle θ is much smaller than 90°, and AC and AB are approximately equal in length, since

$$r = \sqrt{x^2 + y^2} \cong x$$

Referring to Figure 5.8, if we construct a circular segment with radius r and origin A (the dotted line in Fig. 5.8), it is evident that BC is approximately equal to the arc length s; that is,

$$y \cong s$$

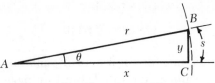

FIGURE 5.8 Right triangle with angle $\angle BAC$ much smaller than 90°.

However, from the discussion in Section 4.6,

$$s = r\theta$$

where θ is expressed in radians. Therefore, we conclude that

$$y \cong r\theta$$

Combining the above results with the defining equations for $\sin \theta$, $\cos \theta$, and $\tan \theta$, we find, for small angles θ,

$$\sin \theta = \frac{y}{r} \cong \frac{r\theta}{r} = \theta$$

$$\cos \theta = \frac{x}{r} \cong \frac{r}{r} = 1$$

$$\tan \theta = \frac{y}{x} \cong \frac{r\theta}{r} = \theta$$

In summary,

For small angles θ,

$$\sin \theta \cong \theta$$

$$\cos \theta \cong 1 \qquad\qquad (5.29)$$

$$\tan \theta \cong \theta$$

where θ is expressed in radians

As an example, for $\theta = 2°$, we first convert to radians and find (to *three* significant figures)

$$\theta = \frac{2}{360} \times 2\pi = 0.0349 \text{ radians}$$

From Equations 5.29 this gives

$$\sin 2° \cong 0.0349$$

$$\cos 2° \cong 1$$

$$\tan 2° \cong 0.0349$$

These approximate values for sin 2°, cos 2°, and tan 2° are in excellent agreement with the values given in Table I on page 193.

When the angle θ is sufficiently large, the approximations in Equations 5.29 are no longer very accurate. For example, $\theta = 15°$ corresponds to

$$\theta = \frac{15}{360} \times 2\pi = 0.262 \text{ radians}$$

The approximations in Equations 5.29 then give

$$\sin 15° \cong 0.262$$

$$\cos 15° \cong 1$$

$$\tan 15° \cong 0.262$$

However, from the trigonometric tables on page 196, the actual values of sin 15°, cos 15°, and tan 15° are (to three significant figures)

$$\sin 15° = 0.259$$

$$\cos 15° = 0.966$$

$$\tan 15° = 0.268$$

Again, remember that approximate calculations are quite sufficient for many purposes. If your answer needs to be no more accurate than 3.5 per cent, then the small-angle approximation for the sine, cosine, and tangent functions for angles less than 15° is entirely satisfactory. For all *precision* calculations, trigonometric tables should be consulted.

EXERCISES

Using Equations 5.29, evaluate

1. $\sin 3° \cong$ _____   (Ans. 132)

2. $\tan 6° \cong$ _____   (Ans. 296)

3. $\sin 3°5 \cong$ _____   (Ans. 316)

4. $\tan 3°6 \cong$ _____   (Ans.   17)

5. $\sin 4°3 \cong$ _____   (Ans. 197)

Compare your answers with the corresponding values of sin θ and tan θ given in Table I, page 193.

CHAPTER 6

FUNCTIONS AND GRAPHS

6.1 CONCEPT OF A FUNCTION

In physics we are frequently called upon to find the *functional relationship* between different physical *variables*. For example, suppose that a particle moves along the X-axis as illustrated in Figure 6.1. We denote by x the particle's displacement from the origin O at the time t. As indicated in Figure 6.1, at time t_1 the particle's displacement is x_1. At some later time, t_2, the particle's displacement is x_2, and so on. That is, the particle's displacement (x) varies as a *function* of the time (t) which has elapsed. Mathematically, we describe this situation by writing

$$x = f(t) \tag{6.1}$$

Equation 6.1 simply states that "x is a function of t." The words "function of" are represented by the symbol $f(\ )$. For each value of t we can determine a corresponding value of x by measuring the particle's displacement from the origin. In Equation 6.1, the quantities x and t are referred to as *variables*. The variable t to which we first assign numerical values (t_1, t_2, etc.) is called the *independent variable*. The variable x, for which we determine the corresponding values, x_1, x_2, etc., is referred to as the *dependent variable*. That is, the displacement x *depends on* the time t which has elapsed.

We note from Figure 6.1 and Equation 6.1 that $x_1 = f(t_1)$, and $x_2 = f(t_2)$, and so on. The specific *form* of $f(t)$ in Equation 6.1 of course depends on the details of the particular problem. For example, we might have $x = at$, or $x = at + b$, or $x = \frac{1}{2} gt^2$, or even $x = $ constant.

The right-hand sides of these equations represent different *functions* of the time t, and the expression that relates t to x is called a *functional relationship*. In the subsequent sections we will examine different types of functional relationships between a pair of variables.

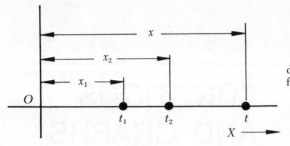

FIGURE 6.1 A particle's displacement (x) varies as a function of time (t).

6.2 REPRESENTATION OF FUNCTIONAL RELATIONSHIPS

The quantitative information collected in an experiment is referred to as *data*. These data are used to determine *functional relationships* between *variables* in the experiment. It is customary to represent these functional relationships by one or more of the following methods:

(1) By means of a *table*.

(2) By means of a *graph*.

(3) By means of an *equation*.

(4) By means of a *verbal statement* which is (preferably) precise.

We now illustrate each of these methods by means of an example.

Suppose that the particle in Figure 6.1 moves in the positive X-direction with a *constant velocity* v_0 where $v_0 = 15$ m/s. Since the velocity is constant, this implies that during each 1-second time interval the particle traverses a distance of 15 m. As in Section 6.1 we represent the displacement of the particle from the origin O by the variable x. Furthermore, we denote the time elapsed by the variable t. For convience, we take $x = 0$ at $t = 0$, by choosing the origin ($x = 0$) to coincide with the location of the particle at the time that we call $t = 0$.

The various methods of representing the functional relationship between the variables x and t for this example are as follows:

(1) Representation by Means of a Table

Since the particle traverses 15 meters in each 1-second time interval, if we measure the displacement x of the particle we will find

At the end of 1 second ($t = 1$ s), $x = 15$ m

At the end of 2 seconds ($t = 2$ s), $x = 15$ m $+ 15$ m $= 30$ m

At the end of 3 seconds ($t = 3$ s), $x = 15$ m $+ 15$ m $+ 15$ m $= 45$ m

and so on. In fact, at the end of N seconds ($t = N$ s), we will find $x = 15N$ meters. The values of t and the corresponding values of x are conveniently summarized (up to $t = 7$ s) in Table 6.1.

TABLE 6.1 THE FUNCTIONAL RELATIONSHIP
BETWEEN THE VARIABLES x AND t

Time t (seconds)	Displacement x (meters)
0	0
1	15
2	30
3	45
4	60
5	75
6	90
7	105

It is important to note that both the *symbols* (t and x) representing the variables and the *units* (seconds and meters, respectively) of these variables are clearly indicated in the table; this procedure should always be followed when tabulating data.

(2) Representation by Means of a Graph

The data which are tabulated in Table 6.1 can also be represented graphically, as shown in Figure 6.2. First, we draw two mutually perpendicular, intersecting axes. By convention, the independent variable (t) is plotted along the horizontal axis (the *abscissa*), whereas the dependent variable (x) is plotted along the vertical axis (the *ordinate*).

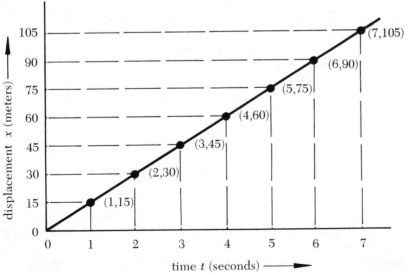

FIGURE 6.2 Graph representing the functional relationship between the variables x and t. Notice that each point has been labeled with coordinates as listed in Table 6.1.

In Figure 6.2 we plot the data of Table 6.1; we note that convenient *scales* have been chosen for the plotting of these data. Specifically, the horizontal axis has been divided into 1-second time intervals, and the vertical axis has been divided into 15-meter length intervals. Note also that both axes in Figure 6.2 have been *clearly labeled* with the appropriate variables (t and x) and units (seconds and meters, respectively). Again, this procedure should always be followed when graphically displaying data.

The plane in which the two axes lie in Figure 6.2 is called the *t-x plane. Any* point in the *t-x* plane can be specified by an ordered pair of numbers (t,x) which are referred to as the *coordinates* of the point (see also Section 4.2). The convention is that the first member of the pair (t) is the *abscissa*, and the second member of the pair (x) is the *ordinate*. (Notice that x is not *always* chosen for the horizontal coordinate.)

To plot the data points which are tabulated in Table 6.1, we proceed as follows:

At $t = 0$ the value of x is $x = 0$. The coordinates of this point are $(0,0)$, which is simply the *origin* in Figure 6.2.

According to Table 6.1, at $t = 1$ s the value of x is $x = 15$ m. The coordinates of this point are $(1,15)$. As indicated by the first pair of dotted lines in Figure 6.2, the point $(1,15)$ is located 1 s along the horizontal axis, and 15 m along the vertical axis.

From Table 6.1, at $t = 2$ s the value of x is $x = 30$ m. The coordinates of this point are $(2,30)$. As indicated by the second pair of dotted lines in Figure 6.2, the point $(2,30)$ is located 2 s along the horizontal axis, and 30 m along the vertical axis.

In a similar manner, the data points (see Table 6.1) $(3,45)$, $(4,60)$, $(5,75)$, $(6,90)$, and $(7,105)$ are plotted in Figure 6.2.

In Figure 6.2 we have connected the data points by a solid line. Since a *single straight line* passes through *all* data points, the functional relationship between the variables x and t in Figure 6.2 is known as a *linear relationship*. For this example the displacement x *increases linearly* with time t at a *constant* rate of 15 m/s.

This example serves to illustrate the *general* method for representing the functional relationship between variables by means of a graph. Whatever the problem, we emphasize the importance of always labeling each coordinate axis with the appropriate *variables* and *units*.

(3) Representation by Means of an Equation

We continue with the same example to illustrate the representation of the functional relationship by means of an equation. Since the displacement x increases with time t at a constant rate of 15 m/s, it is clear that we can relate the variables x and t by the equation

$$x = v_0 t \qquad \text{(with } v_0 = 15 \text{ m/s)} \qquad (6.2)$$

In Equation 6.2 the units of x and t are *meters* and *seconds*, respectively. Furthermore, x is the *dependent variable*, t is the *independent variable*, and $v_0 = 15$ m/s is a *constant*. Equation 6.2 is the equation for a straight line, and is a special example of the general functional relationship shown in Equation 6.1, with $f(t) = v_0 t$.

It is clear that Equation 6.2 can be used to reproduce both Table 6.1 and the graph in Figure 6.2. Since $v_0 = 15$ m/s, it follows from Equation 6.2 that

At $t = 0$, $x = 0$

At $t = 1$ s, $x = (15$ m/s$) \times (1$ s$) = 15$ m

At $t = 2$ s, $x = (15$ m/s$) \times (2$ s$) = 30$ m

At $t = 3$ s, $x = (15$ m/s$) \times (3$ s$) = 45$ m

and so on. These values of t and x are in agreement with the values given in Table 6.1 and in Figure 6.2.

(4) Representation by Means of a Verbal Statement

The functional relationship between the variables x and t in Table 6.1, Figure 6.2, and Equation 6.2 can also be represented by the precise verbal statement,

"The particle's displacement x (in meters) increases with time t (in seconds) at a constant rate of 15 m/s."

For completeness we further qualify this statement by the initial datum that $x = 0$ at $t = 0$; that is, the particle's *initial location* coincides with the origin O in Figure 6.1.

In concluding this section it should be pointed out that there are many physical phenomena for which the functional relationship between variables cannot be described by *precise* verbal statements or equations. A case in point is illustrated in Figure 6.3 where are shown the successive positions of a smoke particle suspended in air, recorded at 1-minute time intervals. The erratic motion of the smoke particle is due to random collisions which the particle makes with the air molecules. Such behavior is generally characteristic of particles undergoing *Brownian motion*. It is clear from the figure that the location of the smoke particle relative to the origin O can be tabulated or plotted graphically for each successive 1-minute time interval. (In fact, Figure 6.3 is just a graph of position with each dot representing the lapse of 1 second.) However, such random motion cannot be represented by a precise verbal statement or equation.

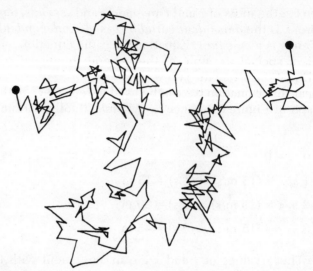

FIGURE 6.3   The successive positions of a smoke particle suspended in air, recorded at 1-minute time intervals.

EXERCISES

1. From the following table construct a graph of
   x (in kilometers) versus t (in seconds).

Distance x (kilometers)	Time t (seconds)
0	0
100	1
200	2
300	3
400	4
500	5

2. What is the equation which relates the vari-
   ables x and t in Exercise 1 above?   (Ans. 106)

3. What is the precise verbal statement which
   relates the variables x and t in Exercise 1
   above?   (Ans. 289)

4. The displacement x (in feet) as a function of
   time t (in seconds) is given by the equation
   $x = v_0 t$, where $v_0 = 55$ ft/s. Present the values
   of x and t from $t = 0$ to $t = 5$ s (in time in-
   intervals of 0.5 s) in tabular form.   (Ans. 86)

5. Construct a graph which relates the variables
   x and t in Exercise 4 above.

6. What is the precise verbal statement which re-
lates the variables x and t in Exercise 4 above?   (Ans. 116)

7. Draw graphs of the following functions from
$t = 0$ to $t = 10$ s:

(a) $x = t$

(b) $x = 2t$

(c) $x = 3t$

(d) $x = 4t$

(e) $x = 5t$

In each case x has units of meters, and t has units of seconds.

6.3 THE GENERAL STRAIGHT LINE

One of the simplest functional relationships that can exist between two variables occurs when one of the variables is directly proportional to the other. Such a functional relationship is said to be a *linear* relationship and can be represented graphically by a *straight line*. The equation

$$x = v_0 t \tag{6.3}$$

considered in Section 6.2 is an example of a straight line.

In this section we consider the general case for a straight line represented by the equation

$$y = ax + b \tag{6.4}$$

In contrast to Equation 6.3, the variable x in Equation 6.4 is the *independent variable*, and the variable y is the *dependent variable*. The quantities a and b in Equation 6.4 are *constants*. We do not specify the units of y, a, x, and b. However, since $y = ax + b$ is an *equation*, it should be noted that y, ax, and b, have the *same* units.

Equation 6.4 is the most general type of *linear* relationship. In the particular case that $b = 0$, y is proportional to x.

The graphical representation of the straight line in Equation 6.4 is illustrated in Figures 6.4 and 6.5 for the case in which a is not equal to zero ($a \neq 0$). The independent variable x is plotted along the horizontal axis (the X-axis), and the dependent variable y is plotted along the vertical axis (the Y-axis). Now, from Equation 6.4 it follows that

for $x = 0$, $y = a \cdot 0 + b = b$

for $y = 0$, $0 = ax + b$, so that $x = -b/a$

That is, the straight line crosses the Y-axis (the line $x = 0$) at $y = b$, and

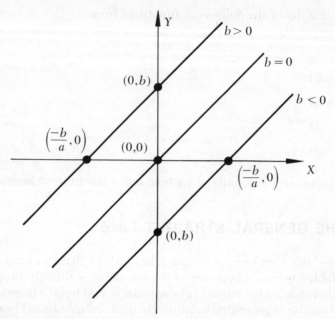

FIGURE 6.4 Graphs of $y = ax + b$ for $a > 0$ and for $b > 0$, $b = 0$, and $b < 0$.

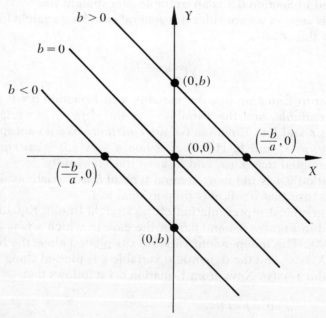

FIGURE 6.5 Graphs of $y = ax + b$ for $a < 0$ and for $b > 0$, $b = 0$, and $b < 0$.

the line crosses the X-axis ($y = 0$) at $x = -b/a$. These two points, $(0,b)$ and $(-b/a,0)$, are called the *intercepts* of the straight line; that is, the line intercepts the axes at these two points. Notice that a and b can be either *positive* or *negative* numbers.

The distinction between Figures 6.4 and 6.5 is that a is positive ($a > 0$) in Figure 6.4 and a is negative ($a < 0$) in Figure 6.5. Notice that lines are drawn for the three possible cases for b: $b > 0$, $b = 0$, and $b < 0$.

In Figure 6.4 (the top line) we have $a > 0$ and $b > 0$. Therefore, $-b/a$ is a *negative* quantity ($-b/a < 0$) and the straight line intercepts the X-axis to the *left* of the origin (i.e., at a *negative* value of x). Since $b > 0$, the line intercepts the Y-axis *above* the origin (i.e., at a *positive* value of y).

For each of the straight lines in Figures 6.4 and 6.5, note whether $-b/a$ is a positive or a negative quantity and similarly for b. Then, determine where the line will intercept each of the axes.

Example 6.3.1

Draw a graph of the straight line $y = 2x + 4$.

For this example, $a = 2$ and $b = 4$. In addition, when $x = 0$, $y = 2 \cdot 0 + 4 = 4$. Therefore, the straight line $y = 2x + 4$ intersects the *positive* Y-axis at the point $(0,4)$. Furthermore, when $y = 0$, $2x + 4 = 0$; that is, $x = -4/2 = -2$. Therefore, the straight line $y = 2x + 4$ intersects the *negative* X-axis at the point $(-2,0)$. Thus, the straight line $y = 2x + 4$ has the form shown in the figure below and is similar to the general straight line represented in Figure 6.4.

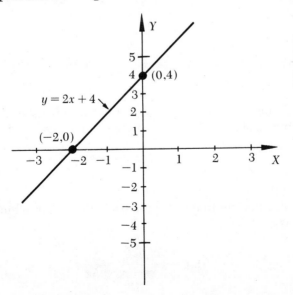

Example 6.3.2

Construct a graph of the equation $y = -2x$.

The equation $y = -2x$ is a special form of the general equation $y = ax + b$ with $b = 0$ and $a = -2$. In this case, if $x = 0$, then $y = 0$. That is, the straight line $y = -2x$ passes through the origin. In order to define the line, we need one additional point. If we choose $x = 1$, then the corresponding value of y is $y = (-2) \cdot 1 = -2$. Therefore, as shown in the figure below, the straight line $y = -2x$ passes through the point $(1, -2)$ as well as the origin $(0, 0)$.

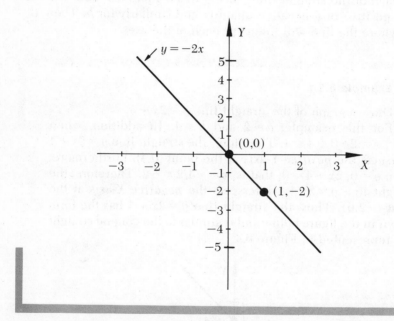

An important characteristic of a straight line is its *slope* or inclination. Referring to Figure 6.6, we now consider a general straight line

$$y = ax + b$$

which passes through the two points (x_1, y_1) and (x_2, y_2). By definition, the *slope* of the straight line in Figure 6.6 is the rate at which y changes with x; that is,

$$\text{Slope} = \frac{y_2 - y_1}{x_2 - x_1} = \frac{\Delta y}{\Delta x} \qquad (6.5)$$

The slope can also be expressed directly in terms of the constant a. Since

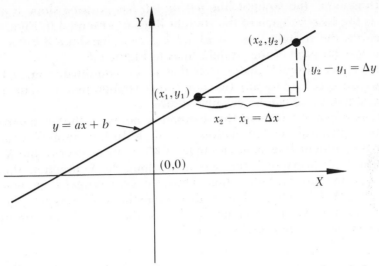

FIGURE 6.6 Graph of the straight line $y = ax + b$ which passes through the points (x_1, y_1) and (x_2, y_2).

the points (x_1, y_1) and (x_2, y_2) are located on the straight line $y = ax + b$, we conclude that

$$y_1 = ax_1 + b \qquad (6.6)$$

and

$$y_2 = ax_2 + b \qquad (6.7)$$

Subtracting Equation 6.6 from Equation 6.7, we find

$$y_2 - y_1 = a(x_2 - x_1)$$

Dividing through by $x_2 - x_1$ gives[*]

$$a = \frac{y_2 - y_1}{x_2 - x_1} \qquad (6.8)$$

Combining Equations 6.5 and 6.8, we can summarize the situation as follows:

The slope of the straight line $y = ax + b$ is

$$\text{Slope} = \frac{y_2 - y_1}{x_2 - x_1} = \frac{\Delta y}{\Delta x} = a \qquad (6.9)$$

Therefore, the slope of the straight line $y = 2x + 4$ (Example 6.3.1) is equal to 2, and the slope of the straight line $y = -2x$ (Example 6.3.2) is equal to -2.

---

[*]Of course, this is possible only if $x_2 - x_1 \neq 0$.

In general, the straight line $y = ax + b$ has *positive* slope if $a > 0$. This is the case for each of the straight lines represented in Figure 6.4. Moreover, the straight line $y = ax + b$ has *negative* slope if $a < 0$. This is the case for each of the straight lines in Figure 6.5.

It is clear from Equation 6.5 that if the coordinates (x_1, y_1) and (x_2, y_2) are specified for *any* two points on a straight line, then the slope of the straight line can be calculated directly.

It is important to note the distinction between the "geometrical" and the "physical" interpretation of a slope. In the graph in Example 6.3.1, the straight line is at an angle of 45° with respect to the X-axis. Therefore, for every inch that we move along the X-direction, the line moves an inch in the Y-direction: Thus, the *geometrical* slope is *unity*. But notice that the physical units attached to the two scales are different. For every unit we move along the X-axis, the line moves upward by *two* units, and the *physical* slope is 2.

Example 6.3.3

What is the slope of a straight line passing through the points (1,2) and (11,8)?

Using the two points given, we have

$$x_1 = 1, \ y_1 = 2$$

and

$$x_2 = 11, \ y_2 = 8$$

Therefore, from Equation 6.5,

$$\text{Slope} = \frac{y_2 - y_1}{x_2 - x_1} = \frac{8 - 2}{11 - 1} = \frac{6}{10} = 0.6$$

It is important to note that if we reverse the order of co-ordinate *labels*, then the value of the slope does not change. That is, if

$$x_1 = 11, \ y_1 = 8$$

and

$$x_2 = 1, \ y_2 = 2$$

then

$$\text{Slope} = \frac{y_2 - y_1}{x_2 - x_1} = \frac{2 - 8}{1 - 11} = \frac{-6}{-10} = \frac{6}{10} = 0.6$$

We now consider the special case in which a straight line has *zero slope*. In this case $a = 0$ and the equation $y = ax + b$ becomes

$$y = b \qquad (6.10)$$

That is, the value of y is equal to the constant b and is the same for *all values of x*. Equation 6.10 is an example of a *constant* functional relationship. The quantity y does not depend on the variable x and so y exhibits no change when the value of x changes. The straight line represented by Equation 6.10 is shown in Figure 6.7. Note that this line is parallel to the X-axis, and intersects the Y-axis at the point $(0,b)$. We reiterate that the straight line $y = b$ has *zero* slope, since $a = 0$ (see Equation 6.9).

Finally, we conclude this section by obtaining an expression for the constant b (Equation 6.4) in terms of the coordinates of any two points, (x_1,y_1) and (x_2,y_2), on the general straight line $y = ax + b$. Eliminating a from Equation 6.6 by means of Equation 6.8, we find

$$y_1 = \left(\frac{y_2 - y_1}{x_2 - x_1}\right) x_1 + b$$

which gives

$$b = y_1 - \left(\frac{y_2 - y_1}{x_2 - x_1}\right) x_1 = \frac{y_1(x_2 - x_1) - (y_2 - y_1)x_1}{x_2 - x_1}$$

This expression for b further reduces to

$$b = \frac{y_1 x_2 - y_2 x_1}{x_2 - x_1} \qquad (6.11)$$

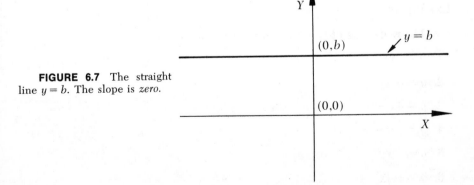

FIGURE 6.7 The straight line $y = b$. The slope is *zero*.

Equations 6.8 and 6.11 express a and b directly in terms of x_1, y_1, x_2, and y_2. We therefore reach the important conclusion that the equation of the straight line $y = ax + b$ which passes through any two points, (x_1, y_1) and (x_2, y_2), can be written as

$$y = \left(\frac{y_2 - y_1}{x_2 - x_1}\right) x + \frac{y_1 x_2 - y_2 x_1}{x_2 - x_1}$$ (6.12)

Example 6.3.4

What is the equation of the straight line passing through the points (0,4) and (16,8)?
We take

$$x_1 = 0, \ y_1 = 4$$

and

$$x_2 = 16, \ y_2 = 8$$

Therefore, from Equation 6.8,

$$a = \frac{y_2 - y_1}{x_2 - x_1} = \frac{8 - 4}{16 - 0} = \frac{4}{16} = \frac{1}{4} = 0.25$$

Furthermore, from Equation 6.11,

$$b = \frac{y_1 x_2 - y_2 x_1}{x_2 - x_1} = \frac{4 \cdot 16 - 8 \cdot 0}{16 - 0} = \frac{64}{16} = 4$$

Thus, the equation of the straight line passing through the points (0,4) and (16,8) is

$$y = 0.25x + 4$$

EXERCISES

Draw graphs of the following straight lines:

1. $y = 3x + 5$

2. $y = 3x$

3. $y = 4x - 6$

4. $y = -x + 2$

5. $y = -x$

6. $y = -4x - 6$

What are the slopes of the following straight lines?

7. $y = 11x + 1$ (Ans. 142)

8. $y = 11x$ (Ans. 107)

9. $y = -x + 5$ (Ans. 326)

10. $y = -6x + 3$ (Ans. 55)

11. $y = 14x + 10$ (Ans. 175)

12. $y = 10^6 x - 1$ (Ans. 218)

Determine the slopes of the straight lines which pass through the following points:

13. $(1,6)$ and $(15,14)$ (Ans. 128)

14. $(0,0)$ and $(3,5)$ (Ans. 201)

15. $(8,9)$ and $(-1,-7)$ (Ans. 139)

16. $(25,24)$ and $(24,25)$ (Ans. 148)

17. $(10,12)$ and $(-12,-10)$ (Ans. 151)

18. $(3,3)$ and $(-156,3)$ (Ans. 205)

Determine the equations of the straight lines which pass through the following points:

19. $(1,6)$ and $(15,14)$ (Ans. 40)

20. $(0,0)$ and $(3,5)$ (Ans. 225)

21. $(8,9)$ and $(-1,-7)$ (Ans. 314)

22. $(25,24)$ and $(24,25)$ (Ans. 299)

23. $(10,12)$ and $(-12,-10)$ (Ans. 217)

24. $(3,3)$ and $(-156,3)$ (Ans. 259)

6.4 FUNCTIONAL RELATIONSHIPS FOR UNIFORMLY ACCELERATED MOTION

In this section we summarize the functional relationships which describe the motion of a particle of mass m along a straight line. As in Figure 6.1, we assume that the particle's motion is along the X-axis and that the particle experiences a *constant* force F_0,

$$F_0 = \text{constant}$$

in the *positive* x-direction. That is, we take $F_0 > 0$. From Newton's Second Law, this constant force F_0 produces a *constant* acceleration a of the particle. The acceleration (see Section 1.9) is determined from

$$\text{Force} = (\text{mass}) \times (\text{acceleration})$$

or,

$$F_0 = ma \tag{6.13}$$

Solving Equation 6.13 for a gives

$$a = \frac{F_0}{m} = \text{constant} \tag{6.14}$$

Note that, since $F_0 > 0$ by hypothesis, the acceleration given by Equation 6.14 is in the *positive* x-direction. Since the acceleration $a = F_0/m$ is constant, the motion produced by the force F_0 is said to be *uniformly accelerated motion*.

We assume that the particle's location at $t = 0$ coincides with the origin O in Figure 6.1; that is,

$$\text{at } t = 0, \ x = 0$$

Further, the initial velocity of the particle is taken to be v_0 (a constant); that is,

$$\text{at } t = 0, \ v = v_0$$

It is one of the results of the study of particle kinematics that the instantaneous velocity v of the particle at time t is

$$v = v_0 + at \tag{6.15}$$

where a is the acceleration (Equation 6.14). In addition, the displacement x of the particle at time t is (if $x = 0$ at $t = 0$)

$$x = v_0 t + \frac{1}{2} at^2 \tag{6.16}$$

Note that Equations 6.15 and 6.16 reduce to the correct results at $t = 0$; that is, $v = v_0$ at $t = 0$, and $x = 0$ at $t = 0$. Equations 6.14 to 6.16 are the *functional relationships* in *equation form* for the uniformly accelerated motion of a particle of mass m which experiences a constant force F_0.

The units for all the quantities used above are given in Section 1.9; see particularly Tables 1.3 and 1.4. It must be emphasized that when applying any equation, such as Equations 6.13 to 6.16 above, *one system of units* (e.g., CGS or MKS) must be used consistently.

We now obtain the *graphical representations* of the equations for uniformly accelerated motion. That is, we construct graphs of

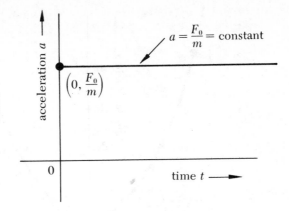

FIGURE 6.8 Graphical representation of acceleration a versus time t for uniformly accelerated motion (Equation 6.14).

(1) acceleration a versus time t (from Eq. 6.14),
(2) velocity v versus time t (from Eq. 6.15), and
(3) displacement x versus time t (from Eq. 6.16).

The acceleration versus time and velocity versus time graphs are illustrated in Figures 6.8 and 6.9, respectively, for times greater than or equal to zero ($t \geq 0$).

The acceleration versus time curve in Figure 6.8 is an example of a *constant functional relationship*. Furthermore, the straight line $a = F_0/m$ intersects the positive acceleration axis, since F_0 is positive (by hypothesis).

In Figure 6.9 we note that the velocity versus time curve is a *straight line with slope equal to the acceleration a*. The straight line $v = v_0 + at$ intersects the velocity axis at $(0, v_0)$. If $v_0 > 0$ (which corresponds to initial motion to the *right* in Fig. 6.1), then $v = v_0 + at$ intersects the *positive* velocity axis as shown in Figure 6.9. However, if $v_0 < 0$ (which corresponds to initial motion to the *left* in Fig. 6.1), then the straight line $v = v_0 + at$ intersects the *negative* velocity axis. In either case the slope of the straight line is *positive*, since $a > 0$.

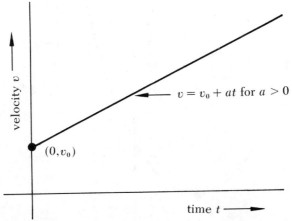

FIGURE 6.9 Graphical representation of velocity v versus time t for uniformly accelerated motion (Equation 6.15).

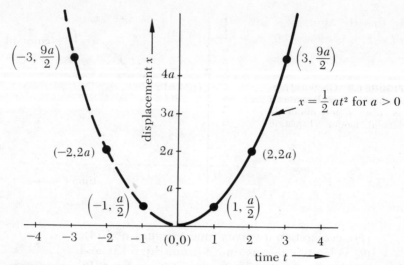

FIGURE 6.10 Graphical representation of displacement x versus time t for uniformly accelerated motion with $v_0 = 0$ and $a > 0$ (Equation 6.17).

We now discuss the graphical representation of Equation 6.16, which relates the variables x and t. When $a \neq 0$, Equation 6.16 is the equation for a *parabola*. For simplicity, we first consider the case in which the initial velocity is zero; that is, $v_0 = 0$. Substituting $v_0 = 0$ into Equation 6.16 gives

$$x = \frac{1}{2}at^2, \text{ when } v_0 = 0 \tag{6.17}$$

The graphical representation of Equation 6.17 is illustrated in Figure 6.10.

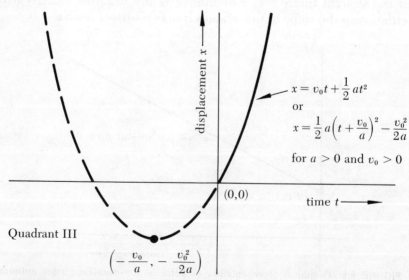

FIGURE 6.11 Graphical representation of displacement x versus time t for uniformly accelerated motion with $v_0 > 0$ and $a > 0$ (Equations 6.18 or 6.16).

Note that the units of x and t have *not* been specified in Figure 6.10. For $t = 1$, it is clear from Equation 6.17 that

$$x = \frac{1}{2} a \cdot (1)^2 = \frac{a}{2}$$

Hence, the point $(1, a/2)$ is located on the parabola in Figure 6.10. We leave it as an exercise for the reader to verify that $(2, 2a)$, $(3, 9a/2)$, $(-1, a/2)$, $(-2, 2a)$, and $(-3, 9a/2)$ are also located on the parabola whose equation is $x = \frac{1}{2} at^2$.

The functional relationship $x = \frac{1}{2} at^2$ has been plotted for negative values of t (the dashed curve in Fig. 6.10), as well as for positive values of t (the solid curve in Fig. 6.10), to illustrate the fact that the function $\frac{1}{2} at^2$ is an *even* (symmetric) function of t. That is,

$$\frac{1}{2} a(-t)^2 = \frac{1}{2} at^2$$

and x has the same value for $-t$ as for $+t$. For our present purposes, however, only positive values of t ($t > 0$) are of *physical* interest. For $t > 0$ we note from Equation 6.17 and Figure 6.10 that the increase in the displacement x is *proportional to* t^2. Furthermore, x increases in a *positive* sense, since $a = F_0/m > 0$.

For the case in which $v_0 \neq 0$, the graphical representation of x versus t can be obtained by first rewriting Equation 6.16 in the equivalent form

$$x = \frac{1}{2} a\left(t + \frac{v_0}{a}\right)^2 - \frac{v_0^2}{2a} \qquad (6.18)$$

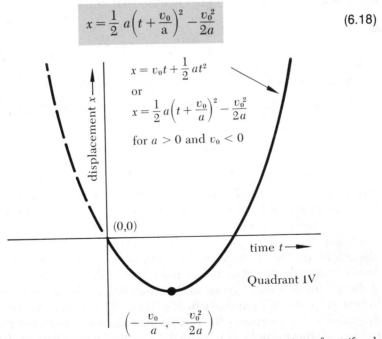

FIGURE 6.12 Graphical representation of displacement x versus time t for uniformly accelerated motion with $v_0 < 0$ and $a > 0$ (Equations 6.18 or 6.16).

Note that the right-hand side of Equation 6.18 is *identical* to the right-hand side of Equation 6.16, since

$$\frac{1}{2} a\left(t + \frac{v_0}{a}\right)^2 - \frac{v_0^2}{2a} = \frac{1}{2} a\left(t^2 + 2t\,\frac{v_0}{a} + \frac{v_0^2}{a^2}\right) - \frac{v_0^2}{2a}$$

$$= \frac{1}{2} at^2 + v_0 t + \frac{v_0^2}{2a} - \frac{v_0^2}{2a}$$

For $a > 0$, it can be shown that the *lowest* point on the parabola represented by Equation 6.18 has the coordinates

$$t = -\frac{v_0}{a}, \text{ and } x = -\frac{v_0^2}{2a} \tag{6.19}$$

Of course, if $v_0 = 0$, then the lowest point corresponds to $(0,0)$ as indicated in Figure 6.10.

The graphical representation of Equation 6.18 (or Eq. 6.16) is illustrated in Figures 6.11 and 6.12 for the two cases, $v_0 > 0$ and $v_0 < 0$, respectively.

Referring to Figure 6.11, we note that when $v_0 > 0$ (that is, when the initial velocity is directed along the *positive* X-axis in Fig. 6.1) the particle's displacement x *increases* in a positive sense for times $t \geq 0$ (the solid curve in Fig. 6.11). Furthermore, the function $x = v_0 t + \frac{1}{2} at^2$ takes on its *minimum* value in Quadrant III when

$$t = -\frac{v_0}{a} < 0, \text{ and } x = -\frac{v_0^2}{2a} < 0 \tag{6.20}$$

Referring to Figure 6.12, we note that when $v_0 < 0$, the function $x = v_0 t + \frac{1}{2} at^2$ takes on its minimum value in Quadrant IV when

$$t = -\frac{v_0}{a} > 0, \text{ and } x = -\frac{v_0^2}{2a} < 0 \tag{6.21}$$

When $v_0 < 0$, the initial particle velocity is directed along the *negative* X-axis in Figure 6.1, which is *opposite* to the direction of the constant force F_0. (Keep in mind that $F_0 = ma > 0$ in the present calculation.) Therefore, for times $t \geq 0$ the particle first undergoes a *negative* displacement ($x < 0$) as indicated in Figure 6.12. The positive force F_0 finally brings the particle to rest and reverses the direction of its motion at the values of x and t given in Equation 6.21. (This corresponds to the lowest point on the parabola in Fig. 6.12.) Subsequently, the particle's displacement x increases for further increases in the time. That is, for $t \geq -v_0/a$, the particle moves in the *positive* x-direction in Figure 6.1.

Example 6.4.1

A block of mass $m = 1$ kg is initially at rest ($v_0 = 0$) on a horizontal frictionless surface. A constant force $F_0 = 10\ N$ is applied to the block in the positive x-direction.

  (a) What is the acceleration of the block?

  (b) What is the velocity of the block after 10 s?

  (c) What is the displacement of the block after 10 s?

In answer to part (a) we find, from Equation 6.14 (see also Table 6.2),

$$\text{acceleration} = a = \frac{F_0}{m} = \frac{10\ \text{kg} \cdot \dfrac{m}{s^2}}{1\ \text{kg}} = 10\ \frac{m}{s^2}$$

In answer to part (b) we find, from Equation 6.15,

$$\text{velocity} = v = v_0 + at = 0 + 10t = 10t \quad \text{(MKS units)}$$

Therefore, for $t = 10$ s,

$$v = \left(10\ \frac{m}{s^2}\right) \cdot (10\ \text{s}) = 100\ \frac{m}{s}$$

In answer to part (c) we find, from Equation 6.16,

$$\text{displacement} = x = v_0 t + \frac{1}{2}at^2 = 0 \cdot t + \frac{1}{2} \cdot 10t^2 = 5t^2 \text{ (MKS units)}$$

Therefore, for $t = 10$ s,

$$x = \left(5\ \frac{m}{s^2}\right) \cdot (10\ \text{s})^2 = 500\ m$$

Example 6.4.2

As illustrated in the figure below, a man on the top of a building 49 m high leans over the edge of the building and drops a ball from rest ($v_0 = 0$). Determine

  (a) the displacement of the ball at 1-second time intervals after the ball leaves the man's hand,

  (b) the velocity of the ball at 1-second time intervals after the ball leaves the man's hand, and

  (c) the time required for the ball to reach the pavement.

Take the acceleration due to gravity (g) to be exactly 9.8 m/s².

As indicated in the figure, we take the positive X-axis to be directed downward along the trajectory of the ball, with the origin O ($x = 0$) coinciding with the point at which the ball is released. Note that positive values of x and v are in the *downward* direction.

Since $a = g = 9.8$ m/s², and $v_0 = 0$, Equations 6.15 and 6.16 can be expressed as

$$\text{velocity} = v = 0 + 9.8t = 9.8t \quad \text{(MKS units)}$$

$$\text{displacement} = x = 0 + \frac{9.8}{2} t^2 = 4.9t^2 \quad \text{(MKS units)}$$

Therefore, in order to answer parts (a) and (b), we construct the following table:

Time t (Seconds)	Displacement x (meters)	Velocity v (m/s)
0	0	0
1	4.9	9.8
2	19.6	19.6
3	44.1	29.4

In order to answer part (c), we solve (for t) the equation, $x = 4.9t^2$, when $x = 49$ m (the height of the building). This gives

$$49 = 4.9t^2$$

or,

$$10 = t^2$$

Therefore, the ball reaches the pavement in $t = \sqrt{10}\,s = 3.162\,s$.

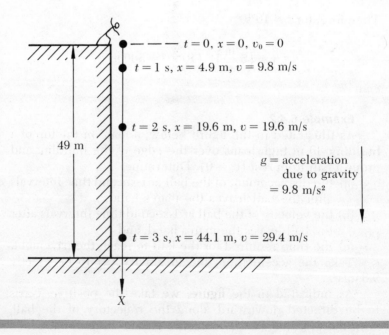

Example 6.4.3

As illustrated in the figure below, a man on the top of a building 49 m high leans over the edge of the building and throws a ball vertically upward with an initial velocity $v_0 = -19.6$ m/s (the *negative* sign occurs since the initial motion is along the *negative* X-axis, as shown in the figure).

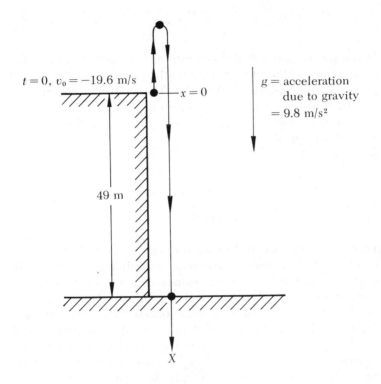

Determine
  (a) the time required for the ball to reach maximum height above the top of the building,
  (b) the maximum height to which the ball rises, and
  (c) the time required for the ball to strike the pavement.
  Take the acceleration due to gravity (g) to be exactly 9.8 m/s².
  As in Example 6.4.2 we take the positive X-direction to be *downward*, with the origin O ($x=0$) coinciding with the point at which the ball is released.
  Since $a = g = 9.8$ m/s², and $v_0 = -19.6$ m/s, Equations 6.15 and 6.16 can be expressed as

$$\text{velocity} = v = -19.6 + 9.8t \quad \text{(MKS units)}$$

$$\text{displacement} = x = -19.6t + 4.9t^2 \quad \text{(MKS units)}$$

In answer to part (a), when the ball reaches the highest point on its trajectory, its velocity is equal to zero. This occurs at a time t determined from

$$v = 0 = -19.6 + 9.8t$$

Solving for t, we find that the ball reaches its maximum height at

$$t = \frac{19.6}{9.8} = 2 \text{ s}$$

It should be noted that this value of t is equal to $-v_0/a$ (Eq. 6.21), which is the *lowest* point on the parabola in Figure 6.12 (because the positive X-direction is *upward* in this figure).

In answer to part (b), to determine the maximum height to which the ball rises, we substitute $t = 2$ s into the equation $x = -19.6t + 4.9t^2$. This gives

$$x = -(19.6) \cdot 2 + 4.9 \cdot (2)^2$$

$$= -39.2 + 19.6$$

$$= -19.6 \text{ m}$$

Therefore, we conclude that the ball rises to a maximum height of 19.6 m above the top of the building. (It should be kept in mind that *negative* values of x correspond to displacements *above* the top of the building.) Notice that the displacement $x = -19.6$ m corresponds to $x = -v_0^2/2a$ (Eq. 6.21), which is the *lowest* point on the parabola in Figure 6.12.

In answer to part (c), we solve (for t) the equation, $x = -19.6t + 4.9t^2$, when $x = 49$ m (the *net* displacement of the ball when it strikes the pavement). This gives

$$49 = -19.6t + 4.9t^2$$

or

$$t^2 - 4t - 10 = 0$$

The solution to this quadratic equation for t is given by (see Section 3.4)

$$t = \frac{4 \pm \sqrt{4^2 + 40}}{2}$$

We discard the solution with a negative value for t (just as in Example 3.4.3), and find

$$t = 2 + \sqrt{14} = 5.742 \text{ s}$$

EXERCISES

1. For Example 6.4.1, draw a graph of acceleration versus time from $t = 0$ to $t = 5$ s, at 1-second time intervals.

2. For Example 6.4.1, draw a graph of velocity versus time from $t = 0$ to $t = 5$ s, at 1-second time intervals.

3. For Example 6.4.1, draw a graph of displacement versus time from $t = 0$ to $t = 5$ s, at 1-second time intervals.

4. For Example 6.4.2, draw a graph of displacement versus time from $t = 0$ to $t = 3$ s, at 1/2-second time intervals.

5. For Example 6.4.3, draw a graph of displacement versus time from $t = 0$ to $t = 5$ s, at 1-second time intervals.

6. For Example 6.4.3, draw a graph of velocity versus time from $t = 0$ to $t = 5$ s, at 1-second time intervals.

7. What is the velocity with which the ball strikes the pavement in Example 6.4.2? (Ans. 9)

8. What is the velocity with which the ball strikes the pavement in Example 6.4.3? (Ans. 313)

9. A block of mass $m = 500$ g moves on a horizontal frictionless surface. The initial velocity of the block is $v_0 = -10^3$ cm/s, and it experiences a constant force, $F_0 = +10^4$ dynes. How long does it take for the force F_0 to bring the block to rest? (Ans. 181)

10. A man on the top of a building 100 m high leans over the edge of the building and releases a ball from rest ($v_0 = 0$). How long does it take for the ball to strike the pavement? Take the acceleration due to gravity (g) to be exactly 9.8 m/s². (Ans. 119)

11. In Exercise 10, what is the velocity of the ball at the instant it strikes the pavement? (Ans. 211)

12. In Exercise 10, what is the displacement of the ball 1 second after it is released? (Ans. 66)

13. A man on the top of a building 49 m high leans over the edge of the building and throws a ball directly downward with a

velocity $v_0 = +19.6$ m/s. How long does it take for the ball to strike the pavement? Take the acceleration due to gravity (g) to be exactly 9.8 m/s².

(Ans. 344)

14. In Exercise 13, what is the velocity of the ball 1 second after it is released?

(Ans. 158)

15. In Exercise 13, what is the velocity of the ball at the instant it strikes the pavement?

(Ans. 233)

6.5 GRAPHS OF SINES AND COSINES

As we saw in Chapter 5, the trigonometric functions, sin θ and cos θ, vary between -1 and $+1$, depending on the value of θ. It is instructive to examine these functions in graphical form. Figure 6.13 shows sin θ and cos θ for θ between 0° and 360°. Notice that these two curves have exactly the same form. If we move the sin θ curve 90° to the left, we have exactly the cos θ curve. That is,

$$\cos(\theta - 90°) = \sin \theta$$
$$\sin(\theta + 90°) = \cos \theta$$

(6.22)

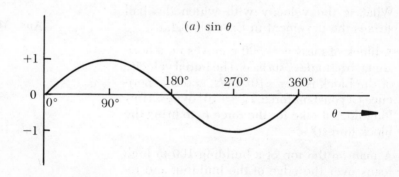

(a) sin θ

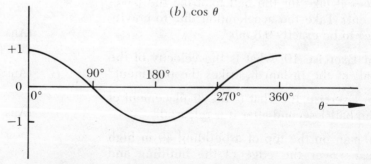

(b) cos θ

FIGURE 6.13   (a) Sin θ and (b) cos θ for θ between 0° and 360°.

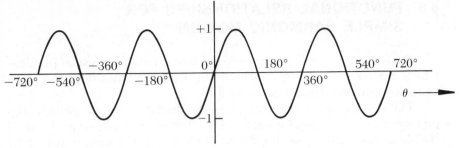

FIGURE 6.14 Sin θ for θ between $-720°$ and $+720°$.

The trigonometric functions are also defined for angles outside the range 0° to 360°. The angle $\theta = -45°$, for example, is the same as the angle $\theta = 315°$ (see Figure 5.7). If we start at $\theta = 0°$ and describe an angle by performing two complete revolutions, we have an angle of 720° which is the same as 360° or 0°. That is, the trigonometric functions repeat every 360°, as shown in Figure 6.14 for sin θ. The corresponding curve for cos θ can be obtained by shifting the sin θ curve by 90° in accordance with Equations 6.22.

In physics problems it is much more usual (and useful) to express θ in radians instead of degrees. In Section 4.6 we showed that $360° = 2\pi$ rad, $180° = \pi$ rad, $90° = \pi/2$ rad, etc. Figure 6.15 illustrates the sine function again, with θ given in radians.

Using radian measure, we can re-express Equations 6.22 as

$$\cos\left(\theta - \frac{\pi}{2}\right) = \sin\theta$$

$$\sin\left(\theta + \frac{\pi}{2}\right) = \cos\theta$$

(6.23)

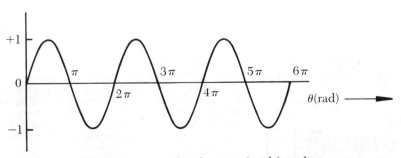

FIGURE 6.15 Sin θ for θ between 0 and 6π radians.

6.6 FUNCTIONAL RELATIONSHIPS FOR SIMPLE HARMONIC MOTION

We next consider an extremely important type of physical problem which involves the appearance of functional relationships between variables that are *trigonometric* in form.

The oscillatory motion of a mass attached to a coiled spring is the prototype of a large and important class of oscillatory phenomena called *simple harmonic motion*. Figure 6.16a shows such a mass at rest in its equilibrium position on a frictionless surface. If we apply an external force to displace the mass to the right, and then suddenly remove the external force, there will be a restoring force F exerted on the mass by the spring and directed to the left (Figure 6.16b).

In general for any displacement x, the restoring force is

$$F = -kx \qquad (6.24)$$

where the *negative* sign in Equation 6.24 means that the direction of the restoring force is *opposite* to the direction of the displacement. In Equation 6.24, k is a *constant* (known as the *force constant*) which is characteristic of the particular spring. We know from Newton's Second Law that the net force F on an object must equal the product of its mass m and its acceleration a. Therefore, at any extension x,

$$F = ma = -kx \qquad (6.25)$$

Solving Equation 6.25 for the acceleration gives

$$a = -\frac{k}{m}x \qquad (6.26)$$

Equation 6.26 is the functional relationship (also known as the *equation of motion*) between the acceleration a and displacement x of an object undergoing *simple harmonic motion*.

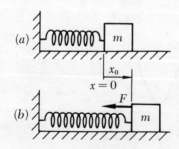

(a)

x_0

$x = 0$

F

(b)

FIGURE 6.16 (a) A mass m rests on a frictionless surface and is attached to a spring; the mass and spring combination is in its normal (equilibrium) condition. (b) If m is displaced to the right an amount x_0 (by an external force), there will be a restoring force to the left, due to the extended spring, given by $F = -kx_0$, where k is the force constant characteristic of the particular spring.

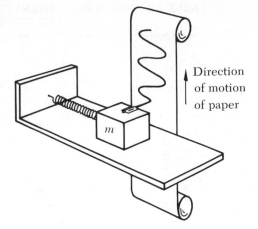

FIGURE 6.17 A simple method for recording the motion of an oscillating mass as a function of time.

Direction
of motion
of paper

If the mass is released from its position of initial extension $(x = x_0)$, in Figure 6.16b, it will be accelerated to the left by the restoring force. At $x = x_0$ the acceleration is a maximum; as the mass moves toward $x = 0$ the velocity increases while the acceleration decreases. When the mass reaches $x = 0$, the restoring force (and, hence, the acceleration) will have decreased to zero, but the velocity of the mass will have increased to its maximum value at this point and the inertia of the mass will carry it into the region of negative x (to the *left* of $x = 0$). In this region the restoring force (and, hence, the acceleration) is directed to the *right* and the mass will be slowed down. At $x = -x_0$, the motion will stop and the acceleration (which is still toward the right) will cause the mass to reverse its motion and move toward $x = x_0$ again. The entire process is one of *cyclic* (or *oscillatory* or *periodic*) motion, with the mass vibrating back and forth between $x = x_0$ and $x = -x_0$.

We can obtain a record of the motion of the mass as a function of time in the following simple way: As shown in Figure 6.17, we attach a pen to the mass and allow it to touch a roll of paper that is moved uniformly in a direction perpendicular to the direction of motion of the mass. In this way we obtain a displacement-time graph of the motion. Examination of the graph in Figure 6.17 shows that it is a *cosine* curve[*] of the form

$$x = x_0 \cos \frac{2\pi}{\tau} t \qquad (6.27)$$

where x_0 is the *amplitude* of the motion (the maximum excursion of the mass from its equilibrium position), and τ is the period of the motion. Notice that in this expression, $(2\pi/\tau)t$ plays the role of the angle θ; as t increases, θ increases. (Remember, *radian* measure is always used in such expressions.) After every interval of time τ the motion repeats

---

[*]Or a *sine* curve, depending on which point we decide to designate as $t = 0$.

TABLE 6.2   THE DISPLACEMENT OF AN OSCILLATING MASS
AS A FUNCTION OF TIME

Time t	$\dfrac{2\pi}{\tau}\, t$	$\cos\dfrac{2\pi}{\tau}\, t$	Displacement $x = x_0 \cos\dfrac{2\pi}{\tau}\, t$
0	0 or 0°	1	x_0
$\dfrac{1}{4}\tau$	$\dfrac{1}{2}\pi$ or 90°	0	0
$\dfrac{1}{2}\tau$	π or 180°	-1	$-x_0$
$\dfrac{3}{4}\tau$	$\dfrac{3}{2}\pi$ or 270°	0	0
τ	2π or 360°	1	x_0

itself indefinitely.* The period τ can be expressed directly in terms of the mass m and force constant k as

$$\tau = 2\pi \sqrt{\frac{m}{k}} \qquad (6.28)$$

For times t between 0 and τ, the functional relationship between x and t (Eq. 6.27) is tabulated in Table 6.2 at successive time intervals of $\tau/4$. To evaluate $\cos 2\pi t/\tau$ for values of $2\pi t/\tau$ greater than 90°, we have made use of the formula derived in Section 5.5. For example, from Equation 5.24,

$$\cos 180° = -\cos(180° - 180°) = -\cos 0° = -1$$

as tabulated in Table 6.2.

The graphical representation of Equation 6.27 is illustrated in Figure 6.18. We note in this figure that the cosine function (and similarly, the sine function) varies in a simple and *regular* way, that is, the variation is *harmonic*. Therefore, the motion described by such functions is called *simple harmonic motion*. It is apparent from Figure 6.18 that the motion repeats itself after every interval of time τ (the period τ is equal to the time interval between any two successive corresponding points on the curve in Figure 6.18).

For a mass undergoing simple harmonic motion according to Equation 6.27, it can be shown that the instantaneous velocity v as a function of time is described by the equation

---

*This is the *ideal* case in which frictional losses are ignored. In a real experiment, friction will be present and the motion will not persist forever unless energy is continually supplied to the system.

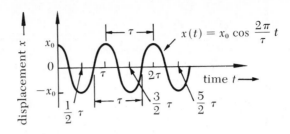

FIGURE 6.18 Graphical representation of the displacement x as a function of time t for a mass undergoing simple harmonic motion (Eq. 6.27).

$$v = -v_m \sin \frac{2\pi}{\tau} t \qquad (6.29)$$

where the maximum velocity v_m is

$$v_m = \frac{2\pi x_0}{\tau}$$

Furthermore, combining Equations 6.26 and 6.27, the acceleration a of the mass can be expressed as

$$a = -a_0 \cos \frac{2\pi}{\tau} t \qquad (6.30)$$

where

$$a_0 = \frac{kx_0}{m}$$

The graphical representation of Equations 6.27, 6.29, and 6.30 is illustrated in Figures 6.19a to 6.19c, respectively.

Figure 6.19a shows the displacement versus time curve (for one complete period τ of oscillation) and is the same as Figure 6.18.

Figure 6.19b shows the velocity versus time curve. At $t = 0$, the velocity $v = 0$, and the displacement x is a maximum ($x = x_0$ from Fig. 6.19a). At $t = \tau/4$, the velocity takes on its maximum negative value $-v_m$. At this instant the mass is passing through $x = 0$ (that is, $x = 0$ at $t = \tau/4$, as evident from Fig. 6.19a). At $t = \tau/2$, the velocity is zero again, and the mass has reached the furthest point to the left ($x = -x_0$ from Fig. 6.19a). For times between $t = \tau/2$ and $t = \tau$, the motion is again to the right with positive velocity. The maximum positive velocity, v_m, is realized at $t = 3\tau/4$. It is evident from Figure 6.19b that the velocity is again zero at $t = \tau$. For times t greater than τ the velocity versus time curve continues to repeat itself in a simple harmonic manner.

Figure 6.19c shows the acceleration versus time curve for a mass undergoing simple harmonic motion. This curve is proportional to the negative of the cosine curve shown in Figure 6.19a. This follows directly from Equations 6.27 and 6.30, and the fact that $a = -kx/m$ (Eq. 6.26).

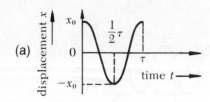

(a)

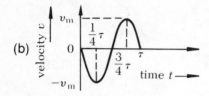

(b)

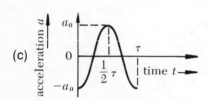

(c)

FIGURE 6.19 Graphical representation of functional relationships for an oscillating mass: (a) displacement versus time (Eq. 6.27), (b) velocity versus time (Eq. 6.29), and (c) acceleration versus time (Eq. 6.30).

Example 6.6.1

A block of mass $m = 10$ g is attached to a spring with force constant* $k = 360$ g/s² (see Fig. 6.16). At $t = 0$ the displacement of the block is $x_0 = +2$ cm, and its velocity is equal to zero.

(a) What is the period τ of the simple harmonic motion?
(b) What is the equation which relates displacement x and time t?
(c) What is the equation which relates velocity v and time t?
(d) What is the equation which relates acceleration a and time t?

In answer to part (a) we find, from Equation 6.28,

$$\text{period} = \tau = 2\pi \sqrt{\frac{m}{k}} = 2\pi \sqrt{\frac{10}{360}} = \frac{2\pi}{6} = \frac{\pi}{3} \text{ s}$$

---

*From Equation 6.25 it is evident that the force constant k has dimensions

$$\frac{\text{mass} \times \text{acceleration}}{\text{displacement}}$$

In CGS units, the dimensions of k are g/s².

Therefore,

$$\tau = 1.05 \text{ s.}$$

In answer to part (b), since $x_0 = +2$ cm, and $\dfrac{2\pi}{\tau} = 6 \text{ s}^{-1}$, we find, from Equation 6.27,

$$\text{Displacement} = x = x_0 \cos \frac{2\pi}{\tau} t = 2 \cos 6t$$

where t is measured in seconds.

In answer to part (c) we find, from Equation 6.29,

$$v_m = \frac{2\pi}{\tau} x_0 = (6 \text{ s}^{-1}) \cdot (2 \text{ cm}) = 12 \frac{\text{cm}}{\text{s}}$$

Therefore, the velocity of the block as a function of time can be expressed as (see Eq. 6.29)

$$\text{velocity} = v = -12 \sin 6t \frac{\text{cm}}{\text{s}}$$

In answer to part (d) we find, from Equation 6.30,

$$a_0 = \frac{k}{m} x_0 = \left(\frac{360 \text{ g/s}^2}{10 \text{ g}}\right) \cdot (2 \text{ cm}) = 72 \frac{\text{cm}}{\text{s}^2}$$

Therefore, the acceleration as a function of time can be expressed as

$$\text{acceleration} = a = -72 \cos 6t \frac{\text{cm}}{\text{s}^2}$$

EXERCISES

1. For Example 6.6.1, draw a graph of displacement versus time from $t = 0$ to $t = \pi/3$ s.

2. For Example 6.6.1, draw a graph of velocity versus time from $t = 0$ to $t = \pi/3$ s.

3. For Example 6.6.1, draw a graph of acceleration versus time from $t = 0$ to $t = \pi/3$ s.

4. A block of mass $m = 10$ g is attached to a spring with force constant $k = 4000$ g/s^2 (see

Fig. 6.16). At $t = 0$ the displacement of the block is $x_0 = +1$ cm, and its velocity is equal to zero. What is the period τ of the simple harmonic motion?

(Ans. 164)

5. In Exercise 4, what is the equation which relates x and t?

(Ans. 290)

6. In Exercise 4, what is the equation which relates v and t?

(Ans. 256)

7. In Exercise 4, what is the equation which relates a and t?

(Ans. 105)

CHAPTER 7

VECTORS

7.1 EXAMPLES OF SCALARS AND VECTORS

Quantities which have *magnitude* only are called *scalars*. Some examples of scalar quantities which occur in physics, together with the symbols that are most frequently used in representing these scalars, are summarized in Table 7.1.

Some physical quantities have *direction* as well as *magnitude*. These quantities are referred to as *vectors*. Some examples of vector quantities, together with the symbols that are most frequently used in representing these vectors, are summarized in Table 7.2. It is customary to place an arrow ($\rightarrow$) above the symbol representing a vector quantity; this is to emphasize that the vector has *direction* as well as magnitude. One also frequently finds vector quantities indicated by bold face type: $\vec{x} = \mathbf{x}$.

To dramatize the distinction between scalars and vectors, consider the illustration in Figure 7.1. A ball is thrown vertically upward. It rises to a height of 20 ft, then falls and is caught at the same location from which it was released. The *net distance* that the ball travels (a *scalar*

TABLE 7.1 EXAMPLES OF SCALARS

Scalar Quantity	Symbol
Distance	x, s
Speed	v
Mass	m
Time	t
Kinetic energy	$\frac{1}{2} mv^2$
Volume	V
Temperature	T

TABLE 7.2 EXAMPLES OF VECTORS

Vector Quantity	Symbol
Displacement	$\vec{x}, \vec{s}$
Velocity	$\vec{v}$
Momentum	$\vec{p} = m\vec{v}$
Force	$\vec{F}$
Acceleration	$\vec{a}$
Electric field	$\vec{E}$
Magnetic field	$\vec{B}$

quantity) is equal to 2×20 ft $= 40$ ft. However, since the initial and final locations of the ball are the same, the *net displacement* of the ball (a *vector* quantity) is *zero*.

7.2 REPRESENTATION AND SIMPLE PROPERTIES OF VECTORS

As shown in Figure 7.2, a vector $\vec{A}$ can be represented pictorially by a *directed line segment* from some origin O (the *foot* of the vector) to a point P (the *head* of the vector). The vector $\vec{A}$ is sometimes denoted by $\overrightarrow{OP}$. The direction of the straight line segment, as inferred by the direction of the arrowhead in Figure 7.2, corresponds to the *direction* of the vector $\vec{A}$. The length of the straight line segment is proportional to the *magnitude* of $\vec{A}$, in some agreed units. It is customary to denote the magnitude of $\vec{A}$ by $|\vec{A}|$, or simply by A. That is,

$$\text{Magnitude of vector } \vec{A} = |\vec{A}| \text{ or } A \tag{7.1}$$

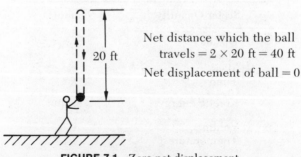

Net distance which the ball
travels $= 2 \times 20$ ft $= 40$ ft

Net displacement of ball $= 0$

20 ft

FIGURE 7.1 Zero net displacement.

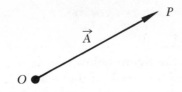

FIGURE 7.2   Pictorial representation of a vector $\vec{A}$.

A vector $\vec{A}$ can be multiplied directly by a scalar b. The resulting quantity is also a vector, which we denote by $\vec{A}'$, where

$$\vec{A}' = b\vec{A} \qquad (7.2)$$

The direction of $\vec{A}'$ is either *parallel* or *antiparallel* to $\vec{A}$, depending on whether b is a positive number or a negative number, respectively. For example, if $b = 2.5$, then

$$\vec{A}' = 2.5\,\vec{A}$$

On the other hand, if $b = -1$, then

$$\vec{A}' = -\vec{A}$$

The vectors $2.5\,\vec{A}$, $-\vec{A}$, together with the vector $\vec{A}$, are shown in Figure 7.3.

The vector $2.5\,\vec{A}$ is parallel to the vector $\vec{A}$; that is, the new vector points in the *same direction* as $\vec{A}$. The vector $-\vec{A}$, however, is antiparallel to $\vec{A}$, since it points in the direction *opposite* to $\vec{A}$. Furthermore, keeping in mind that the magnitude of a vector is always a *positive* number, or zero (if $\vec{A}$ is a *null* vector, $\vec{A} = 0$), we find that the magnitude of the vectors $2.5\,\vec{A}$ and $-\vec{A}$ are, respectively, $2.5\,A$ and A. Referring to Equation 7.2, we find in the general case that the magnitude of the vector $\vec{A}'$ is equal to $|b|\,A$, where $|b|$ is the *magnitude* of the number b; that is,

$$A' = |b|\,A \qquad (7.3)$$

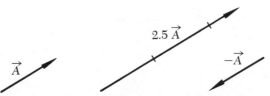

FIGURE 7.3   The vectors $\vec{A}$, $2.5\,\vec{A}$, and $-\vec{A}$. The vector $-\vec{A}$ is antiparallel to the other two vectors.

Example 7.2.1

As shown in the figure below, the vector $\vec{v}$ is a velocity of 50 m/s directed 30° north of due east. What are the magnitudes and directions of the vectors (a) $2\vec{v}$, (b) $-\vec{v}$, and (c) $-2\vec{v}$?

The answers to these questions are summarized pictorially in the figure below.

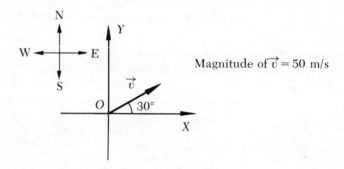

Referring to figure (a), the vector $2\vec{v}$ is in the same direction as $\vec{v}$, and has a magnitude equal to twice the magnitude of $\vec{v}$. Therefore, $2\vec{v}$ is a velocity of 2×50 m/s = 100 m/s directed *30° north of due east.*

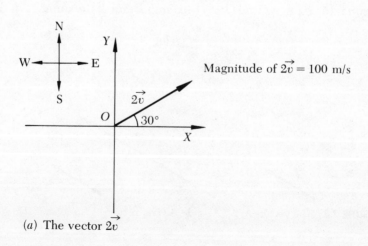

(a) The vector $2\vec{v}$

Referring to figure (b), the vector $-\vec{v}$ is in the direction opposite to $\vec{v}$, and has a magnitude equal to the magnitude of $\vec{v}$. Therefore, $-\vec{v}$ is a velocity of 50 m/s directed *30° south of due west.*

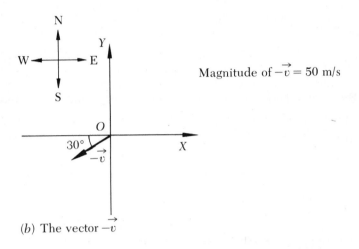

Magnitude of $-\vec{v} = 50$ m/s

(*b*) The vector $-\vec{v}$

Referring to figure (c), the vector $-2\vec{v}$ is also in the direction opposite to $\vec{v}$, but has a magnitude equal to twice the magnitude of $\vec{v}$. Therefore, $-2\vec{v}$ is a velocity of 2×50 m/s$=$ 100 m/s directed *30° south of due west.*

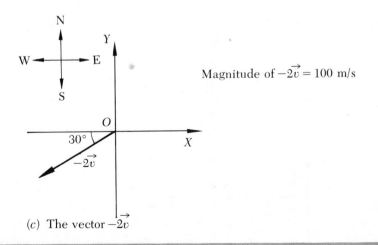

Magnitude of $-2\vec{v} = 100$ m/s

(*c*) The vector $-2\vec{v}$

The vector $\vec{x}$ is a displacement of 25 mi directed 40° north of due west. What are the magnitudes and directions of the following vectors?

1. $6\vec{x}$ (Ans. 303)

2. $-1.5\vec{x}$ (Ans. 301)

3. $-4\vec{x}$ (Ans. 165)

4. $3\vec{x}$ (Ans. 285)

The vector $\vec{F}$ is a force of 14 newtons directed vertically downward on a horizontal floor. What are the magnitudes and directions of the following vectors?

5. $5000\,\vec{F}$ (Ans. 176)

6. $-0.5\,\vec{F}$ (Ans. 84)

7. $-15\,\vec{F}$ (Ans. 134)

8. $13\,\vec{F}$ (Ans. 172)

7.3 ADDITION AND SUBTRACTION OF VECTORS

In a flat portion of rural Kansas, a motorist drives 10 miles due east from point O to point P. His *displacement* during this portion of the journey is represented by the vector $\vec{A} = \overrightarrow{OP}$ in Figure 7.4. At P the motorist turns left and then drives 20 miles due north to point P', his final destination. This displacement is represented by the vector $\vec{B} = \overrightarrow{PP'}$ in Figure 7.4. It is clear that the trip from O to P, followed by the trip from P to P', has the same end result as a trip *directly* from O to P'.

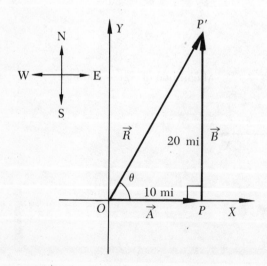

FIGURE 7.4 A journey in Kansas from O to P to P'.

As shown in the figure, the *net* displacement for the entire trip can be represented by the vector $\vec{R} = \overrightarrow{OP'}$.

The vector displacement $\vec{R}$ is called the *resultant* of the two vector displacements, $\vec{A}$ and $\vec{B}$. We can express $\vec{R}$ in equation form as

$$\vec{R} = \vec{A} + \vec{B} \tag{7.4}$$

Equation 7.4 is a *vector equation*, and states that the resultant vector displacement $\vec{R}$ is equal to the sum of the two vector displacements, $\vec{A}$ and $\vec{B}$. We note from Figure 7.4 that the vectors $\vec{A}$, $\vec{B}$, and $\vec{R}$ together form the three sides of the triangle *OPP'*.

The magnitude and direction of the resultant vector $\vec{R}$ can be determined in the following manner: Referring to Figure 7.4, and making use of the fact that *OPP'* is a *right* triangle with angle $\angle OPP' = 90°$, we find, from the Pythagorean theorem,

$R = $ magnitude of resultant vector $\vec{R}$ in Figure 7.4

$= \sqrt{(10)^2 + (20)^2}$ mi $= \sqrt{500}$ mi $= 22.36$ mi

Next, we let θ represent the angle $\angle P'OP$. From the definition of $\tan \theta$,

$$\tan \theta = \frac{20 \text{ mi}}{10 \text{ mi}} = 2$$

We determine θ from page 205 of the Trigonometric Tables, which gives $\theta = 63°.4$.

Therefore, we have found that the resultant vector $\vec{R}$ is equal to a displacement of 22.36 miles directed 63°.4 north of due east. We emphasize that to determine completely the vector $\vec{R}$, *both* its magnitude and direction must be specified.

The above example illustrates a further important property of the vector addition in Equation 7.4. The vector equation, $\vec{R} = \vec{A} + \vec{B}$, does *not* imply that the magnitude of the resultant vector $\vec{R}$ is equal to the sum of the magnitudes of the vectors $\vec{A}$ and $\vec{B}$. In Figure 7.4 we see that $A + B = 10$ mi $+ 20$ mi $= 30$ mi, whereas $R = 22.36$ mi. Therefore, even though $\vec{R} = \vec{A} + \vec{B}$ is true, $R = A + B$, is *not* true. This is, in fact, a general property of vector addition, and is not restricted to the specific example illustrated in Figure 7.4.

The example discussed above also serves to demonstrate the general method for finding the resultant of any two vectors $\vec{A}$ and $\vec{B}$ which are *orthogonal*, that is, *perpendicular* to one another.

We now develop a procedure for determining the resultant $\vec{R}$ of any two vectors $\vec{A}$ and $\vec{B}$ with *arbitrary* orientations. The procedure is known as the *method of triangles*. Consider the vectors $\vec{A}$ and $\vec{B}$ represented pictorially in Figures 7.5(a) and (b). The vectors $\vec{A}$ and $\vec{B}$ can signify displacements, velocities, momenta, forces, or any other type of vector.

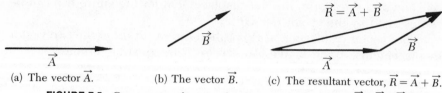

(a) The vector $\vec{A}$.   (b) The vector $\vec{B}$.   (c) The resultant vector, $\vec{R} = \vec{A} + \vec{B}$.

FIGURE 7.5 Constructing the triangle of vectors to determine $\vec{R} = \vec{A} + \vec{B}$.

As shown in Figure 7.5(c), we determine the resultant vector $\vec{R} = \vec{A} + \vec{B}$ by constructing a triangle in which $\vec{A}$ and $\vec{B}$ are adjacent sides and connect head to foot. The resultant vector $\vec{R}$ is then equal to the third side of the triangle. That is, the magnitude of $\vec{R}$ corresponds to the length of the third side, and the direction of $\vec{R}$ is that indicated in Figure 7.5(c). It is important to note that, while $\vec{A}$ and $\vec{B}$ connect head to foot, the direction of $\vec{R}$ is such that $\vec{A}$ and $\vec{R}$ connect foot to foot, and $\vec{B}$ and $\vec{R}$ connect head to head.

It is also important to note that the order in which $\vec{A}$ and $\vec{B}$ are connected does not affect the resultant $\vec{R}$. The two possibilities are illustrated in Figure 7.6. In Figure 7.6, the resultant vector $\vec{R}$ has the same magnitude and direction for both triangles. Therefore, in practice we can construct $\vec{R}$ by either method indicated in Figure 7.6. This important result is a reflection of the fact that vector addition obeys the *commutative law*; that is,

$$\vec{A} + \vec{B} = \vec{B} + \vec{A} \qquad (7.5)$$

The procedure illustrated in Figures 7.5 and 7.6 can be extended to determine the resultant $\vec{R}$ of any number of vectors $\vec{A}, \vec{B}, \vec{C}, \cdots$. For example, to determine the resultant $\vec{R} = \vec{A} + \vec{B} + \vec{C} + \vec{D} + \vec{E}$ for the five vectors $\vec{A}, \vec{B}, \vec{C}, \vec{D}$, and $\vec{E}$, we construct the hexagon (6-sided polygon) shown in Figure 7.7. In this figure the vectors $\vec{A}, \vec{B}, \vec{C}, \vec{D}$, and $\vec{E}$ are connected together, head to foot, and form five adjacent sides of a hexagon. The resultant vector $\vec{R}$ is then equal to the sixth side; that is, the magnitude of $\vec{R}$ corresponds to the length of the sixth side, and the direction of $\vec{R}$ is that indicated in Figure 7.7. Note that $\vec{A}$ and $\vec{R}$ connect foot to foot, whereas $\vec{E}$ and $\vec{R}$ connect head to head. As before, the order in which the vectors $\vec{A}, \vec{B}, \vec{C}, \vec{D}$, and $\vec{E}$ are connected does not affect the magnitude or direction of $\vec{R}$. (Verify this statement by trying several different orderings for adding the vectors.)

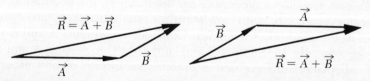

FIGURE 7.6 Two possible triangles for constructing the resultant vector, $\vec{R} = \vec{A} + \vec{B}$.

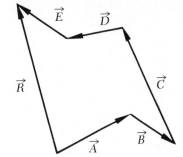

FIGURE 7.7 A hexagon of vectors for constructing the resultant vector, $\vec{R} = \vec{A} + \vec{B} + \vec{C} + \vec{D} + \vec{E}$.

The procedure for *subtracting* a vector $\vec{B}$ from a vector $\vec{A}$ is straightforward. We first reverse the direction of $\vec{B}$, that is, we form the vector $-\vec{B}$, and then *add* it to $\vec{A}$. The difference vector $\vec{D}$ is denoted by

$$\vec{D} = \vec{A} - \vec{B} \qquad (7.6)$$

Since we can write $-\vec{B} = +(-\vec{B})$, Equation 7.6 can also be expressed in the form

$$\vec{D} = \vec{A} + (-\vec{B}) \qquad (7.7)$$

That is, $\vec{D}$ is equal to the *sum* of the vectors $\vec{A}$ and $-\vec{B}$. As illustrated in Figure 7.8, we determine the difference vector, $\vec{D} = \vec{A} - \vec{B}$, by constructing a vector triangle in which $\vec{A}$ and $-\vec{B}$ form adjacent sides. The vector $\vec{D}$ is then equal to the third side of the triangle. That is, the magnitude of $\vec{D}$ corresponds to the length of the third side, and the direction of $\vec{D}$ is that indicated in Figure 7.8. We reiterate that the difference vector, $\vec{D} = \vec{A} - \vec{B}$, is equal to the resultant of the vectors $\vec{A}$ and $-\vec{B}$.

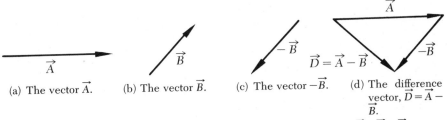

(a) The vector $\vec{A}$. (b) The vector $\vec{B}$. (c) The vector $-\vec{B}$. (d) The difference vector, $\vec{D} = \vec{A} - \vec{B}$.

FIGURE 7.8 Constructing the triangle of vectors to determine $\vec{D} = \vec{A} - \vec{B}$.

EXERCISES

The vector $\vec{v}$ is a velocity of 60 m/s directed 45° north of due east. What are the magnitudes and directions of the resultants of the following vectors?

1. $\vec{v} + 3\vec{v}$ (Ans. 288)

2. $\vec{v} - 3\vec{v}$ (Ans. 212)

3. $\vec{v} + 2\vec{v} + 4\vec{v}$ (Ans. 178)

4. $\vec{v} + 2\vec{v} - 4\vec{v}$ (Ans.   25)

5. $\vec{v} + 2\vec{v} + 3\vec{v} - 6\vec{v}$ (Ans.   39)

6. As shown in the figure, a pilot is attempting to fly his airplane due north with a velocity $\vec{v}$ equal to 300 mi/hr. There is a crosswind with $\vec{v}_{wind}$ equal to 100 mi/hr in the easterly direction. What is the magnitude and direction of the resultant velocity $\vec{v}_R$ of the airplane?   (Ans.   28)

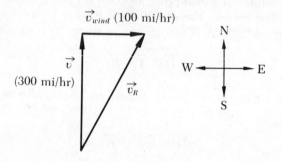

7. If $\vec{F}_1$ is a force of 2 N in the positive X-direction and $\vec{F}_2$ is a force of 2 N in the positive Y-direction (see the figure), what is the magnitude and direction of the resultant force, $\vec{R}_F = \vec{F}_1 + \vec{F}_2$?   (Ans.   14)

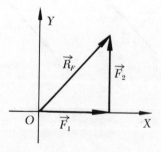

8. Refer to Exercise 7 and replace the force $\vec{F}_2$ by a force $2\vec{F}_2$. What is the magnitude and direction of the new resultant force, $\vec{R}'_F = \vec{F}_1 + 2\vec{F}_2$?   (Ans.   4)

7.4   VECTOR ADDITION BY THE METHOD OF TRIANGLES

In this section we consider an example of vector addition by the method of triangles discussed in the preceding section.

Suppose that the motorist referred to at the beginning of Section 7.3, after traveling 10 mi due east from O to P, then travels 20 mi from P to P' *in a direction 60° north of due east* (rather than 20 mi *due north*). His journey can then be represented by the two displacement vectors $\vec{A}$ and $\vec{B}$ shown in Figure 7.9.

The resultant displacement vector $\vec{R} = \vec{A} + \vec{B}$ forms the third side OP' of the triangle OPP', and has the direction indicated in the figure.

The following construction can be used to determine the magnitude and direction of $\vec{R}$. From the point P' we draw a straight line perpendicular to the X-axis. Note that the X-axis is aligned along the vector $\vec{A}$. In Figure 7.9 the intersection point is denoted by Q, and

$$\text{length of } PQ = x$$

$$\text{length of } QP' = y$$

Evidently, the triangle OQP' is a *right* triangle with angle $\angle OQP' = 90°$. This fact will subsequently allow us to determine the magnitude and direction of $\vec{R}$ by elementary trigonometry.

We first determine the values of x and y. Referring to the right triangle PQP' in Figure 7.9, we have

$$\text{length of } PP' = 20 \text{ mi}$$

$$\text{angle } \angle P'PQ = 60°$$

$$\text{angle } \angle PQP' = 90°$$

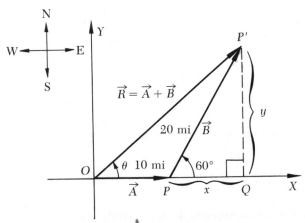

FIGURE 7.9   Another journey in Kansas from O to P to P'.

From the definition of the cosine function, it is apparent from the triangle PQP' that

$$\cos 60° = \frac{x}{20 \text{ mi}}$$

Therefore,

$$x = (20 \text{ mi}) \cos 60° = (20 \text{ mi}) \cdot (0.5) = 10 \text{ mi}$$

Furthermore, from the definition of the sine function,

$$\sin 60° = \frac{y}{20 \text{ mi}}$$

Therefore,

$$y = (20 \text{ mi}) \sin 60° = (20 \text{ mi}) \cdot (0.866) = 17.32 \text{ mi}$$

We can now determine the magnitude and direction of the resultant vector, $\vec{R} = \vec{A} + \vec{B}$. Referring to the right triangle OQP', and making use of the fact that $x = 10$ mi and $y = 17.32$ mi, we note that

$$\text{length of } OQ = A + x = 10 \text{ mi} + 10 \text{ mi} = 20 \text{ mi}$$

$$\text{length of } QP' = y = 17.32 \text{ mi}$$

Therefore, from the Pythagorean theorem,

$$R = \text{magnitude of resultant vector } \vec{R}$$
$$= \sqrt{(20)^2 + (17.32)^2} \text{ mi} = \sqrt{700} \text{ mi} = 26.5 \text{ mi}$$

Furthermore, from the triangle OQP' in Figure 7.9, it is apparent that

$$\tan \theta = \frac{\text{length of } QP'}{\text{length of } OQ} = \frac{17.32 \text{ mi}}{20 \text{ mi}} = 0.866$$

where $\theta = \angle P'OQ$. From the Trigonometric Tables on page 201, $\tan \theta = 0.866$ gives

$$\theta = 40°.9$$

In summary, the resultant vector $\vec{R}$ in Figure 7.9 is equal to a displacement of 26.5 mi in a direction 40°.9 north of due east.

EXERCISES

1. In the figure below, what is the magnitude and direction of the resultant vector, $\vec{R} = \vec{B} + \vec{A}$?

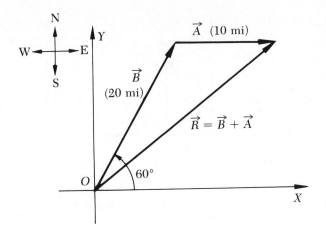

Compare your answer with the example worked out above.

(Ans. 140)

2. In the figure below, what is the magnitude and direction of the resultant vector, $\vec{R} = \vec{A} + \vec{B}$?

(Ans. 360)

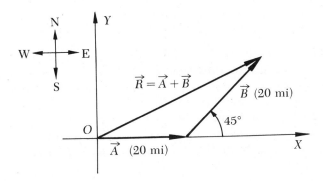

7.5 RESOLUTION OF VECTORS INTO RECTANGULAR COMPONENTS

By using the basic rules of trigonometry, any vector $\vec{A}$ can be *decomposed* into two vectors, the sum of which yields the original vector. This process is especially useful when the two vectors, which are called *component* vectors, are at right angles to one another. In this case, the two component vectors are referred to as *rectangular* component vectors. The general procedure for *resolving* a vector $\vec{A}$ into two rectangular components, $\vec{A}_x$ and $\vec{A}_y$, is illustrated in Figure 7.10.

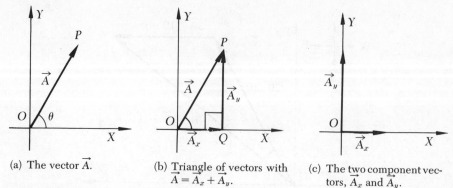

(a) The vector $\vec{A}$.

(b) Triangle of vectors with $\vec{A} = \vec{A}_x + \vec{A}_y$.

(c) The two component vectors, $\vec{A}_x$ and $\vec{A}_y$.

FIGURE 7.10   Resolving a vector $\vec{A}$ into two rectangular components, $\vec{A}_x$ and $\vec{A}_y$: $\vec{A} = \vec{A}_x + \vec{A}_y$.

In Figure 7.10(a), we let θ be the angle between the positive X-axis and the vector $\vec{A}$. In Figure 7.10(b), from the head of vector $\vec{A}$ (that is, the point P) we drop a perpendicular PQ to the X-axis. The angle $\angle OQP$ is then equal to $90°$. It is evident from Figure 7.10(b) that OPQ forms a triangle of vectors consisting of $\vec{A}_x$, $\vec{A}_y$, and $\vec{A}$. Furthermore,

$$\vec{A} = \vec{A}_x + \vec{A}_y \tag{7.8}$$

That is, the vector sum of $\vec{A}_x$ and $\vec{A}_y$ is equal to the vector $\vec{A}$. The original vector $\vec{A}$ in Figure 7.10(a) can therefore be *removed* and *replaced* by the two rectangular components, $\vec{A}_x$ and $\vec{A}_y$. We emphasize that the two component vectors, $\vec{A}_x$ and $\vec{A}_y$ in Figure 7.10(c), are *totally equivalent* to the original vector $\vec{A}$ in Figure 7.10(a).

In Figure 7.10(b) the triangle OQP is a right triangle with $\angle OQP = 90°$, and $\angle POQ = \theta$. From the definition of $\cos \theta$, we find

$$\cos \theta = \frac{OQ}{OP} = \frac{A_x}{A}$$

This gives

$$A_x = A \cos \theta \tag{7.9}$$

where A is the magnitude of vector $\vec{A}$. Similarly, from the definition of $\sin \theta$, we find

$$\sin \theta = \frac{PQ}{OP} = \frac{A_y}{A}$$

This gives

$$A_y = A \sin \theta \qquad (7.10)$$

Therefore, if the magnitude A of the original vector $\overrightarrow{A}$ and the angle θ which $\overrightarrow{A}$ makes with the positive X-axis are specified, the rectangular components, A_x and A_y, can be determined from the equations

$$A_x = A \cos \theta$$
$$A_y = A \sin \theta \qquad (7.11)$$

Example 7.5.1

If $\overrightarrow{A}$ represents a force vector of 50 dynes directed at an angle of 60° with respect to the positive X-axis, what are the values of A_x and A_y?
Evidently,

$$A = 50 \text{ dynes, and } \theta = 60°$$

From Equation 7.11 we find

$$A_x = (50 \text{ dynes}) \cos 60° = (50 \text{ dynes}) \cdot (0.5) = 25 \text{ dynes}$$

and

$$A_y = (50 \text{ dynes}) \sin 60° = (50 \text{ dynes}) \cdot (0.866) = 43.3 \text{ dynes}$$

Thus, the original force vector is equivalent to a force of 25 dynes in the positive X-direction, combined with a force of 43.3 dynes in the positive Y-direction.

It should be pointed out that the formulae for A_x and A_y given in Equation 7.11 are also valid when the angle θ is larger than 90°, that is, when the vector $\overrightarrow{A}$ lies in Quadrants II, III, or IV.

EXAMPLE 7.5.2

Find the values of A_x and A_y in the situation illustrated in the accompanying figure.

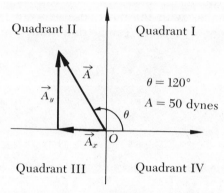

Quadrant II

Quadrant I

$\theta = 120°$

$A = 50$ dynes

Quadrant III

Quadrant IV

Referring to the figure above and to Equation 7.11, we find

$$A_y = (50 \text{ dynes}) \sin 120° \tag{1}$$

However, from Equation 5.24 in Section 5.5, we can write

$$\sin 120° = \sin (180° - 120°) = \sin 60° = 0.866 \tag{2}$$

Therefore,

$$A_y = (50 \text{ dynes}) \cdot (0.866) = 43.3 \text{ dynes} \tag{3}$$

Furthermore, referring again to the figure and to Equation 7.11, we find

$$A_x = (50 \text{ dynes}) \cos 120°$$

But from Equation 5.24 in Section 5.5, we note that the cosine function is negative in Quadrant II, and that

$$\cos 120° = -\cos(180° - 120°) = -\cos 60° = -0.5$$

Therefore,

$$A_x = (50 \text{ dynes}) \cdot (-0.5) = -25 \text{ dynes} \tag{4}$$

The fact that the value of A_x given in (4) is a *negative* number is simply a reflection of the fact that the component vector $\vec{A}_x$ is in the *negative* X-direction. Note that the value of A_y given in (3) is a *positive* number, since the component vector $\vec{A}_y$ is in the *positive* Y-direction.

We conclude this section by noting that a specification of the rectangular components, A_x and A_y, permits a *complete* determination of the vector $\vec{A}$; that is, both the magnitude and the direction of $\vec{A}$ can be obtained directly from a knowledge of A_x and A_y. The procedure is as follows.

If we square each expression in Equation 7.11 and add the resultant expressions, we find

$$A_x^2 + A_y^2 = A^2 \cos^2\theta + A^2 \sin^2\theta$$
$$= A^2(\cos^2\theta + \sin^2\theta) = A^2$$

where we have used the fact that $\cos^2\theta + \sin^2\theta = 1$. Taking the square root then gives

$$A = \sqrt{A_x^2 + A_y^2} \qquad (7.12)$$

Equation 7.12 expresses the magnitude of the vector $\vec{A}$ in terms of the rectangular components, A_x and A_y.

Furthermore, from Equation 7.11 we find

$$\frac{A_y}{A_x} = \frac{A \sin\theta}{A \cos\theta} = \tan\theta$$

That is,

$$\tan\theta = \frac{A_y}{A_x} \qquad (7.13a)$$

or,

$$\theta = \tan^{-1}(A_y/A_x) \qquad (7.13b)$$

Equation 7.13b expresses the angle θ which the vector $\vec{A}$ makes with the positive X-axis (see Fig. 7.10) directly in terms of A_x and A_y.

Therefore, for the given rectangular components, A_x and A_y, Equations 7.12 and 7.13 can be used to determine completely the vector $\vec{A}$. The magnitude A of the vector $\vec{A}$ is obtained from Equation 7.12 and the direction of the vector $\vec{A}$ (in particular, the angle θ which $\vec{A}$ makes with the positive X-axis) is obtained from Equation 7.13.

Example 7.5.3

For $A_x = 3$ cm and $A_y = 4$ cm, what are the values of A and θ?

From Equation 7.12,

$$A = \sqrt{3^2 + 4^2} \text{ cm} = \sqrt{25} \text{ cm} = 5 \text{ cm}$$

and from Equation 7.13,

$$\tan \theta = \frac{4 \text{ cm}}{3 \text{ cm}} = 1.333$$

Making use of the Trigonometric Tables, page 203, we find

$$\theta = 53°1$$

Thus, the vector $\vec{A}$ represents a displacement of 5 cm directed at an angle of 53°1 with respect to the positive X-axis.

EXERCISES

The vector $\vec{A}$ represents a displacement of 5 mi directed along the positive X-axis. Determine the following quantities:

1. $A = $ _____   (Ans. 321)

2. $\theta = $ _____   (Ans. 221)

3. $A_x = $ _____   (Ans. 332)

4. $A_y = $ _____   (Ans. 61)

The vector $\vec{A}$ represents a force of 500 dynes directed along the negative Y-axis. Determine the following quantities:

5. $A = $ _____   (Ans. 18)

6. $\theta = $ _____   (Ans. 188)

7. $A_x = $ _____   (Ans. 122)

8. $A_y = $ _____   (Ans. 222)

The vector $\vec{A}$ represents a velocity of 10^8 cm/s directed at an angle of 18° measured from the positive X-axis. Determine the following quantities:

9. $A = $ _____   (Ans. 238)

10. $\theta = $ _____   (Ans. 90)

11. $A_x = $ _____   (Ans. 147)

12. $A_y = $ ____   (Ans. 352)

The vector $\overrightarrow{A}$ represents a momentum of 10^7 g-cm/s directed at an angle of 135° measured from the positive X-axis. Determine the following quantities:

13. $A = $ ____   (Ans. 230)

14. $\theta = $ ____   (Ans. 63)

15. $A_x = $ ____   (Ans. 271)

16. $A_y = $ ____   (Ans. 12)

The rectangular components of a vector $\overrightarrow{A}$ are $A_x = 10^5$ cm and $A_y = 10^5$ cm.

17. What is the magnitude of $\overrightarrow{A}$?   (Ans. 96)

18. What is the direction of $\overrightarrow{A}$?   (Ans. 231)

The rectangular components of a vector $\overrightarrow{A}$ are $A_x = -10$ newtons and $A_y = 15$ newtons.

19. What is the magnitude of $\overrightarrow{A}$?   (Ans. 257)

20. What is the direction of $\overrightarrow{A}$?   (Ans. 46)

7.6 ADDITION OF VECTORS BY COMPONENTS

The resolution of vectors into rectangular components, as discussed in Section 7.5, provides us with a powerful tool for the addition of vectors. Consider the two vectors, $\overrightarrow{A}$ and $\overrightarrow{B}$, which are to be added to-

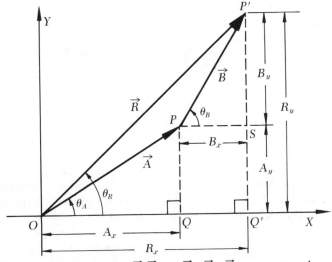

FIGURE 7.11   Resolving the vectors $\overrightarrow{A}$, $\overrightarrow{B}$, and $\overrightarrow{R} = \overrightarrow{A} + \overrightarrow{B}$, into rectangular components.

gether, as shown in Figure 7.11, to form the resultant vector $\vec{R} = \vec{A} + \vec{B}$. The diagram, because it shows in detail the method of constructing the components, appears cluttered. But the process of vector addition by components is, in fact, quite simple. The x-component of the resultant vector is just the sum of the x-components of the two vectors being added; and similarly for the y-components.

In Figure 7.11 the lines PQ and PQ' are drawn perpendicular to the X-axis. Furthermore, the line PS is drawn parallel to the X-axis. The angles $\angle POQ$, $\angle P'OQ'$, and $\angle P'PS$ are denoted by θ_A, θ_R, and θ_B, respectively. Clearly, θ_A is the angle which the vector $\vec{A}$ makes with positive X-axis, and θ_R is the angle which the resultant vector $\vec{R}$ makes with the positive X-axis. Moreover, since PS is parallel to the X-axis, θ_B is the angle which the vector $\vec{B}$ makes with the positive X-axis.

It is evident that A_x, B_x, and R_x, are the rectangular components of the vectors $\vec{A}$, $\vec{B}$, and $\vec{R}$, respectively, along the X-axis. Furthermore, by inspection of Figure 7.11, we find that

$$R_x = A_x + B_x \tag{7.14}$$

Similarly, A_y, B_y, and R_y, the rectangular components of the vectors $\vec{A}$, $\vec{B}$, and $\vec{R}$, respectively, along the Y-axis, are related by

$$R_y = A_y + B_y \tag{7.15}$$

Thus, if the values of A_x, B_x, A_y, and B_y, are known, we can determine the rectangular components, R_x and R_y, of the resultant vector, $\vec{R} = \vec{A} + \vec{B}$, directly from Equations 7.14 and 7.15. Furthermore, making use of Equations 7.14 and 7.15, and applying the Pythagorean theorem to the right triangle $OQ'P'$ in Figure 7.11, we find

$$R = \text{magnitude of resultant vector } \vec{R}$$
$$= \sqrt{R_x^2 + R_y^2}$$
$$= \sqrt{(A_x + B_x)^2 + (A_y + B_y)^2} \tag{7.16}$$

Thus, the magnitude of the resultant vector $\vec{R}$ is given directly in terms of A_x, B_x, A_y, and B_y.

We can also express the angle θ_R, which the resultant vector $\vec{R}$ makes with the X-axis, in terms of A_x, B_x, A_y, and B_y. Referring to the right triangle $OQ'P'$ in Figure 7.11, and making use of Equations 7.14 and 7.15 and the definition of $\tan \theta_R$, we find

$$\tan \theta_R = \frac{R_y}{R_x} = \frac{A_y + B_y}{A_x + B_x}$$

Therefore, in terms of A_x, B_x, A_y, and B_y, the angle θ_R is given by

$$\tan \theta_R = \frac{A_y + B_y}{A_x + B_x} \tag{7.17a}$$

or,

$$\theta_R = \tan^{-1} \left(\frac{A_y + B_y}{A_x + B_x} \right) \tag{7.17b}$$

The usefulness of Equations 7.16 and 7.17 is obvious. Given two vectors $\vec{A}$ and $\vec{B}$, to determine the resultant vector, $\vec{R} = \vec{A} + \vec{B}$, we first resolve the vectors $\vec{A}$ and $\vec{B}$ into their rectangular components. That is, we determine the values of A_x, A_y, B_x, and B_y. Then, substituting these values into Equations 7.16 and 7.17, we obtain the magnitude R of the resultant vector $\vec{R}$ (Eq. 7.16), and the angle θ_R which $\vec{R}$ makes with the X-axis (Eq. 7.17b).

In most applications, it is usually the values of A, θ_A, B, and θ_B (see Fig. 7.11) which are prescribed. That is, the magnitudes and directions of the vectors $\vec{A}$ and $\vec{B}$ are usually given. From Figure 7.11, and the defining equations for the sine and cosine functions (see also Eq. 7.11), we note that

$$A_x = A \cos \theta_A$$
$$A_y = A \sin \theta_A$$
$$B_x = B \cos \theta_B \tag{7.18}$$
$$B_y = B \sin \theta_B$$

Therefore, in Equations 7.16 and 7.17, A_x, A_y, B_x, and B_y, can be eliminated in favor of A, θ_A, B, and θ_B. Substituting Equation 7.18 into Equation 7.16 gives

$$R = \sqrt{(A \cos \theta_A + B \cos \theta_B)^2 + (A \sin \theta_A + B \sin \theta_B)^2} \tag{7.19}$$

for the magnitude R of the resultant vector $\vec{R}$. Furthermore, substituting Equation 7.18 into Equations 7.17, we obtain

$$\tan \theta_R = \frac{A \sin \theta_A + B \sin \theta_B}{A \cos \theta_A + B \cos \theta_B} \tag{7.20a}$$

or,

$$\theta_R = \tan^{-1} \left(\frac{A \sin \theta_A + B \sin \theta_B}{A \cos \theta_A + B \cos \theta_B} \right) \tag{7.20b}$$

which determines the angle θ_R that the resultant vector $\vec{R}$ makes with the X-axis. We emphasize that Equations 7.19 and 7.20 express R and θ_R *directly* in terms of the magnitudes and directions of the vectors $\vec{A}$ and $\vec{B}$, that is, in terms of A, θ_A, B, and θ_B.

Example 7.6.1

To dramatize the usefulness of Equations 7.19 and 7.20, or Equations 7.16 and 7.17, we reexamine the example discussed in Section 7.4.

Comparing Figure 7.9 with Figure 7.11, we find

$$\theta_A = 0°$$
$$A = 10 \text{ mi}$$

and

$$\theta_B = 60°$$
$$B = 20 \text{ mi}$$

Therefore,

$$A_x = A \cos \theta_A = (10 \text{ mi}) \cos 0° = 10 \text{ mi}$$
$$A_y = A \sin \theta_A = (10 \text{ mi}) \sin 0° = 0$$

Similarly,

$$B_x = B \cos \theta_B = (20 \text{ mi}) \cos 60° = (20 \text{ mi}) \cdot 0.5 = 10 \text{ mi}$$
$$B_y = B \sin \theta_B = (20 \text{ mi}) \sin 60° = (20 \text{ mi}) \cdot 0.866 = 17.32 \text{ mi}$$

Substituting into Equation 7.19 or into Equation 7.16, we find that the magnitude of the resultant vector $\vec{R}$ is

$$R = \sqrt{(10 + 10)^2 + (0 + 17.32)^2} \text{ mi} = \sqrt{(20)^2 + (17.32)^2} \text{ mi}$$
$$= \sqrt{700} \text{ mi} = 26.5 \text{ mi}$$

Furthermore, substituting into Equation 7.20 or into Equation 7.17, we find*

$$\tan \theta_R = \frac{0 + 17.32 \text{ mi}}{10 + 10 \text{ mi}} = \frac{17.32}{20} = 0.866$$

---

*For this example, we note that the angle θ_R in Figure 7.11 and the angle θ in Figure 7.9 represent the *same* angle, namely the angle which $\vec{R}$ makes with the X-axis.

From the Trigonometric Tables on page 201, $\tan \theta_R = 0.866$ gives

$$\theta_R = 40^\circ.9$$

The above values for R and θ_R are, of course, identical to the results obtained in Section 7.4 by direct construction. We emphasize, however, the *brevity* of the present calculation, *given* the general formulae in Equations 7.16 and 7.17, or in Equations 7.19 and 7.20.

Figure 7.11, together with the general procedure which leads to Equations 7.16 and 7.17, can readily be extended to determine the magnitude and direction of the vector $\vec{R}$ which results from the addition of any number of vectors $\vec{A}, \vec{B}, \vec{C}, \vec{D}, \cdots$. We summarize here the results for the case in which there are *three* component vectors $\vec{A}, \vec{B}$, and $\vec{C}$. In this case the resultant vector $\vec{R}$ is

$$\vec{R} = \vec{A} + \vec{B} + \vec{C} \tag{7.21}$$

The rectangular component R_x of the vector $\vec{R}$ along the X-axis is

$$R_x = A_x + B_x + C_x \tag{7.22}$$

where A_x, B_x, and C_x are the rectangular components of the vectors $\vec{A}, \vec{B}$, and $\vec{C}$, respectively, along the X-axis. Similarly,

$$R_y = A_y + B_y + C_y \tag{7.23}$$

is the rectangular component of the vector $\vec{R}$ along the Y-axis. The reader is encouraged to extend the constructions in Figure 7.11 to the case in which there are three component vectors $\vec{A}, \vec{B}$, and $\vec{C}$, and reproduce the results given in Equations 7.22 and 7.23.

Equations 7.22 and 7.23 can be combined to give the magnitude R of the resultant vector $\vec{R}$. We find

$$R = \text{magnitude of resultant vector } \vec{R}$$
$$= \sqrt{R_x^2 + R_y^2}$$
$$= \sqrt{(A_x + B_x + C_x)^2 + (A_y + B_y + C_y)^2}$$

If we denote by θ_R the angle which $\vec{R}$ makes with the positive X-axis, then

$$\tan \theta_R = \frac{R_y}{R_x} = \frac{A_y + B_y + C_y}{A_x + B_x + C_x}$$

The above results can be summarized as

$$R = \sqrt{(A_x + B_x + C_x)^2 + (A_y + B_y + C_y)^2} \qquad (7.24)$$

and

$$\theta_R = \tan^{-1}\left(\frac{A_y + B_y + C_y}{A_x + B_x + C_x}\right) \qquad (7.25)$$

Equations 7.24 and 7.25 express the magnitude and direction of the resultant vector $\vec{R}$ *directly* in terms of the rectangular components of the vectors $\vec{A}$, $\vec{B}$, and $\vec{C}$.

It should be noted that Equations 7.24 and 7.25 are quite analogous in structure to Equations 7.16 and 7.17b. In fact, if $\vec{C}$ is the *null* vector (that is, if $\vec{C} = 0$), then $C_x = 0$ and $C_y = 0$, and Equations 7.24 and 7.25 reduce directly to Equations 7.16 and 7.17b, as would be expected.

EXERCISES

All symbols, definitions, and so on, used in the following exercises refer to the configuration illustrated in Figure 7.11.

If $A_x = 1$ cm, $A_y = 3$ cm, $B_x = 5$ cm, and $B_y = 3$ cm, calculate the following quantities:

1. $R_x = $ ____   (Ans. 189)

2. $R_y = $ ____   (Ans.   1)

3. $R = $ ____   (Ans. 129)

4. $\theta_R = $ ____   (Ans.   13)

If $A_x = 10$ cm/s, $A_y = -30$ cm/s, $B_x = 30$ cm/s, and $B_y = 60$ cm/s, calculate the following quantities:

5. $R_x = $ ____   (Ans. 82)

6. $R_y = $ ____   (Ans. 249)

7. $R = $ ____   (Ans. 339)

8. $\theta_R = $ ____   (Ans. 341)

If $A = 14.14$ dynes, $\theta_A = 45°$, $B = 100$ dynes, $\theta_B = 60°$, calculate the following quantities:

9. $A_x = $ ____   (Ans. 27)

10. $A_y = $ ____   (Ans. 52)

11. $B_x = $ ____   (Ans. 265)

12. $B_y = $ ____   (Ans. 161)

13. $R_x = $ ____   (Ans. 15)

14. $R_y =$ ——   (Ans. 325)

15. $R =$ ——   (Ans. 83)

16. $\theta_R =$ ——   (Ans. 180)

If $A = 50$ dynes, $\theta_A = 0°$, $B = 25$ dynes, $\theta_B = 90°$, calculate the following quantities:

17. $A_x =$ ——   (Ans. 48)

18. $A_y =$ ——   (Ans. 117)

19. $B_x =$ ——   (Ans.   7)

20. $B_y =$ ——   (Ans. 346)

21. $R_x =$ ——   (Ans. 350)

22. $R_y =$ ——   (Ans. 281)

23. $R =$ ——   (Ans. 333)

24. $\theta_R =$ ——   (Ans. 127)

7.7 SCALAR PRODUCT OF VECTORS

The multiplication of two scalar quantities, or even the multiplication of a scalar and a vector, is an easy process to understand. But how does one multiply two *vectors*? Because vectors have direction as well as magnitude, vector multiplication must be carefully defined. There are two important and different ways in which two vectors can be multiplied. The first of these is a process that produces a *scalar* result and is called the *scalar product*. The second type of multiplication produces a *vector* result and is called the *vector product* (see Section 7.8).

The scalar product can be understood in terms of the following physical situation. Suppose that a force $\vec{F}$ is applied to an object and as a result the object is displaced from a position A to a position B; this displacement is defined by the vector $\vec{s}$ (see Fig. 7.12). The *work* that is done by the force in causing the displacement is proportional to the component of $\vec{F}$ that is in the direction of the displacement $\vec{s}$. The component of $\vec{F}$ that is *perpendicular* to $\vec{s}$ (in this case, the *vertical* component) causes no displacement because the motion is entirely horizontal; this component of $\vec{F}$ does no work. Therefore, the work done by the force is done entirely by the horizontal component of $\vec{F}$:

work done = (magnitude of displacement $\vec{s}$)

$\times$ (component of force $\vec{F}$ in direction of $\vec{s}$)   (7.26)

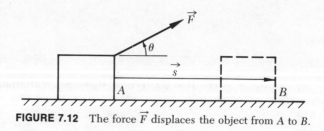

FIGURE 7.12   The force $\vec{F}$ displaces the object from A to B.

If θ is the angle between $\vec{s}$ and $\vec{F}$ (see Fig. 7.12), the component of $\vec{F}$ in the direction of $\vec{s}$ is

$$\text{component of } \vec{F} \text{ in direction of } \vec{s} = F \cos\theta \qquad (7.27)$$

Then, the work done by $\vec{F}$ is expressed as

$$W = Fs \cos\theta \qquad (7.28)$$

This type of multiplication of two vectors ($\vec{F}$ and $\vec{s}$) is indicated in a shorthand notation by

$$W = \vec{F} \cdot \vec{s} \qquad (7.29)$$

The dot between $\vec{F}$ and $\vec{s}$ denotes the *scalar product* (the result W is a *scalar* quantity). Sometimes this product is called the *dot* product. The scalar product of two vectors always means the product of the *magnitudes* of the two vectors multiplied by the cosine of the angle between them. For two general vectors, $\vec{A}$ and $\vec{B}$, as in Figure 7.13, the scalar product is

$$\vec{A} \cdot \vec{B} = |\vec{A}|\,|\vec{B}| \cos\theta = AB \cos\theta \qquad (7.30)$$

The scalar product is independent of the order of multiplication:

$$\vec{A} \cdot \vec{B} = \vec{B} \cdot \vec{A} \qquad (7.31)$$

That is, it does not matter whether we consider the product to be the component of $\vec{A}$ in the direction of $\vec{B}$ multiplied by $|\vec{B}|$ or the component of $\vec{B}$ in the direction of $\vec{A}$ multiplied by $|\vec{A}|$.

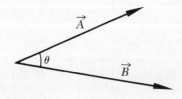

FIGURE 7.13   The scalar product of $\vec{A}$ and $\vec{B}$ is $\vec{A} \cdot \vec{B} = AB \cos\theta$.

Example 7.7.1

If, in Figure 7.12, the force $\vec{F}$ is 50 N in a direction 30° above the horizontal (i.e., $\theta = 30°$) and if the horizontal displacement is 10 m, calculate the work done.

$$W = \vec{F} \cdot \vec{s} = Fs \cos \theta$$

$$= (50 \text{ N}) \times (10 \text{ m}) \times \cos 30°$$

$$= (500 \text{ N-m}) \times 0.866$$

$$= 433 \text{ N-m}$$

But 1 N-m is equal to 1 joule; therefore,

$$W = 433 \text{ J}$$

EXERCISES

1. If $A = 1$ cm, $B = 10$ cm, and $\theta = 60°$, then
   $\vec{A} \cdot \vec{B} =$ _____.   (Ans. 200)

2. If $A = 1$ cm, $B = 10$ cm, and $\theta = 90°$, then
   $\vec{A} \cdot \vec{B} =$ _____.   (Ans. 123)

3. If $F = 500$ dynes, $s = 70$ cm, and $\theta = 0°$, then
   $\vec{F} \cdot \vec{s} =$ _____.   (Ans. 292)

4. If $F = 32$ newtons, $s = 10$ m, and $\theta = 0°$, then
   $\vec{F} \cdot \vec{s} =$ _____.   (Ans. 173)

5. If $F = 500$ dynes, $s = 70$ cm, and $\theta = 20°$, then
   $\vec{F} \cdot \vec{s} =$ _____.   (Ans. 198)

7.8   THE VECTOR PRODUCT

The second type of vector multiplication gives a product which has both *magnitude* and *direction*, and hence is known as the *vector product*.

The vector product can be illustrated by means of the following physical example. Suppose that an object of mass m has instantaneous velocity $\vec{v}$. Its *linear momentum* is then given by the expression $\vec{p} = m\vec{v}$.

Furthermore, suppose that the *displacement* of the object, relative to some origin O (see Fig. 7.14) is $\vec{r}$. As illustrated in Figure 7.14, the two vectors $\vec{r}$ and $\vec{p}$ determine a plane, which we take to be the X-Y plane. In addition, θ denotes the angle between the vectors $\vec{r}$ and $\vec{p}$. The Z-axis, of course, is perpendicular to the X-Y plane.

By definition, the *angular momentum* of the object relative to the origin O is equal to the *vector product* of $\vec{r}$ and $\vec{p}$, which is written as $\vec{r} \times \vec{p}$ and is read as "$\vec{r}$ cross $\vec{p}$." The term "cross" distinguishes the *vector* product from the *scalar* (or dot) product. It is customary to denote $\vec{r} \times \vec{p}$ by the vector $\vec{L}$. That is,

$$\vec{r} \times \vec{p} = \vec{L} \tag{7.32}$$

The vector product $\vec{r} \times \vec{p} = \vec{L}$ represents the angular momentum of the object, and has both magnitude and direction.

We first consider the *direction* of the angular momentum vector $\vec{L}$. As illustrated in Figure 7.15, $\vec{L} = \vec{r} \times \vec{p}$ is *perpendicular* to the plane formed by the vectors $\vec{r}$ and $\vec{p}$. Furthermore, $\vec{L} = \vec{r} \times \vec{p}$ points in the direction in which a *right-hand screw* would advance if the head of the screw is rotated from the first vector ($\vec{r}$) toward the second vector ($\vec{p}$). Therefore, the direction of $\vec{L}$ is along the positive Z-axis in Figure 7.15.

The *magnitude* of the angular momentum vector $\vec{L} = \vec{r} \times \vec{p}$ is defined to be the magnitude of $\vec{r}$ times the magnitude of the component of $\vec{p}$ perpendicular to $\vec{r}$. Referring to Figure 7.14 or Figure 7.15, we note that the component of $\vec{p}$ perpendicular to $\vec{r}$ is

$$p_{\perp} = p \sin \theta$$

where p is the magnitude of the vector $\vec{p}$, and θ is the angle between $\vec{r}$ and $\vec{p}$. Themfore, we find that the magnitude of $\vec{L}$ (which we denote by $|\vec{L}| = L$) is given by

$$L = rp_{\perp} = rp \mid \sin \theta \mid \tag{7.33}$$

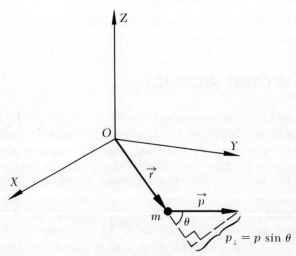

FIGURE 7.14 Vectors representing the displacement $\vec{r}$ and momentum $\vec{p}$ of a particle with mass m.

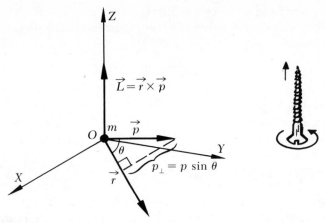

FIGURE 7.15 The vector product $\vec{r} \times \vec{p} = \vec{L}$ represents the angular momentum of the mass relative to the origin O. The vector $\vec{L} = \vec{r} \times \vec{p}$ points in the direction in which a right-hand screw would advance if the head of the screw is rotated from the first vector ($\vec{r}$) toward the second vector ($\vec{p}$) in the product. The magnitude of $\vec{L}$ is equal to $rp|\sin \theta|$.

where r denotes the magnitude of $\vec{r}$, and $|\sin \theta|$ denotes the magnitude of $\sin \theta$. Evidently, $L = 0$ if $\vec{p}$ is *parallel* to $\vec{r}$. Referring to Figure 7.14, we see that $\theta = 0°$ in this case. Therefore, for $\vec{p}$ parallel to $\vec{r}$,

$$L = pr \mid \sin 0° \mid = 0$$

Similarly, if $\vec{p}$ is *antiparallel* to $\vec{r}$ ($\theta = 180°$), we also find

$$L = pr \mid \sin 180° \mid = 0$$

In contrast, if $\vec{p}$ is perpendicular to $\vec{r}$ ($\theta = 90°$), then

$$L = pr \mid \sin 90° \mid = pr$$

and L is a *maximum* for a given p and r if $\theta = 90°$.

Example 7.8.1

If, in Figure 7.14, $\theta = 30°$, $r = 10$ m, and $p = 30$ kg-m/s, what is the magnitude L of the angular momentum?
Since $\theta = 30°$, we find

$$\sin \theta = \sin 30° = 0.5$$

Substituting the values for r, p, and $\sin \theta$ into Equation 7.33 we find

$$L = (10 \text{ m}) \cdot (30 \text{ kg-m/s}) \cdot (0.5)$$
$$= 150 \text{ kg-m}^2/\text{s}$$

The vector product of two arbitrary vectors, $\vec{A}$ and $\vec{B}$, is defined in a completely analogous manner. The magnitude and direction of the vector product

$$\vec{A} \times \vec{B} = \vec{C} \tag{7.34}$$

is illustrated in Figure 7.16.

The general prescription for determining the magnitude and direction of the vector product $\vec{A} \times \vec{B} = \vec{C}$ is summarized below.

Magnitude and direction of the vector product $\vec{A} \times \vec{B} = \vec{C}$:

The *direction* of $\vec{A} \times \vec{B} = \vec{C}$ is the direction which a right-hand screw would advance if the head of the screw is rotated from the first vector ($\vec{A}$) toward the second vector ($\vec{B}$) in the product (see Fig. 7.15). The *magnitude* of $\vec{A} \times \vec{B} = \vec{C}$ is given by

$$C = AB \, |\sin \theta|$$

where θ is the angle between $\vec{A}$ and $\vec{B}$, and A, B, and C denote the magnitudes of the vectors $\vec{A}$, $\vec{B}$, and $\vec{C}$, respectively.

It is important to note that the direction of a vector product *does* depend on the order in which the vectors are multiplied (this is in contrast to the scalar product of two vectors discussed in Section 7.7). Referring to Figure 7.16, it is clear that the direction of $\vec{B} \times \vec{A}$ is *opposite* to the direction of $\vec{A} \times \vec{B}$. This follows, since the direction which a right-hand screw would advance if the head of the screw is rotated from $\vec{B}$ toward $\vec{A}$ (rather than from $\vec{A}$ toward $\vec{B}$) is along $-\vec{C}$ in Figure 7.16. Since the *magnitude* of the vector product remains unchanged by the order of vector multiplication, we therefore conclude

$$\vec{B} \times \vec{A} = -\vec{A} \times \vec{B}$$

or equivalently,

$$\vec{A} \times \vec{B} = -\vec{B} \times \vec{A} \tag{7.35}$$

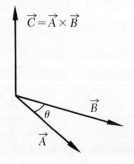

FIGURE 7.16 The vector product $\vec{A} \times \vec{B} = \vec{C}$ points in the direction which a right-hand screw would advance if the head of the screw is rotated from the first vector ($\vec{A}$) toward the second vector ($\vec{B}$) in the product. The magnitude of $\vec{C}$ is equal to $AB \, |\sin \theta|$.

EXERCISES

1. If $\vec{A}$ is directed due east, and $\vec{B}$ is directed due north, then the direction of $\vec{A} \times \vec{B}$ is ____.   (Ans. 263)

2. If $\vec{A}$ is directed due north, and $\vec{B}$ is directed due east, then the direction of $\vec{A} \times \vec{B}$ is ____.   (Ans. 272)

3. If $\vec{A}$ is directed 36°.1 north of due east, and $\vec{B}$ is directed due north, then the direction of $\vec{A} \times \vec{B}$ is ____.   (Ans. 345)

4. If $\vec{A}$ is directed due east, and $\vec{B}$ is directed due south, then the direction of $\vec{A} \times \vec{B}$ is ____.   (Ans. 26)

5. If $\vec{A}$ is directed due east, and $\vec{B}$ is directed due south, then the direction of $\vec{B} \times \vec{A}$ is ____.   (Ans. 154)

6. If $A = 6$ cm, $B = 10$ cm, and $\theta = 45°$, then $C = |\vec{A} \times \vec{B}| = $ ____.   (Ans. 234)

7. If $r = 10^6$ m, $p = 10^5$ kg-m/s, and $\theta = 90°$, then $L = $ ____.   (Ans. 29)

8. If $r = 10^6$ m, $p = 10^5$ kg-m/s, and $\theta = 0°$, then $L = $ ____.   (Ans. 252)

9. If $r = 10^{-8}$ cm, $p = 10^{-18}$ gm-cm/s, and $\theta = 90°$, then $L = $ ____.   (Ans. 223)

10. The distance of the Earth from the Sun is $r = 1.5 \times 10^8$ km, and the mean orbital speed of the Earth is $v = 30$ km/s. The mass of the Earth is 6.0×10^{24} kg, and its orbit about the Sun is approximately circular ($\theta = 90°$ in Fig. 7.14). What is the magnitude of the Earth's angular momentum around the Sun?   (Ans. 88)

CHAPTER

8 THE SLIDE RULE

8.1 INTRODUCTION

In studying physics we wish to concentrate on the *physics* instead of the mathematical detail. Therefore, it is important to make efficient use of the time spent on computational work. For this reason the student is urged to learn how to carry out computations on the slide rule, which is a relatively quick way of obtaining answers to three (and sometimes four) significant figures. The slide rule can be used for

(1) multiplication
(2) division
(3) squares and square roots

as well as for some other operations which we will not study here. The slide rule can *not* be used for addition or subtraction.

This chapter is intended as a *guide* to using the slide rule for the basic operations listed above, and does not constitute a complete course on the slide rule. The material presented here should be supplemented by a thorough reading of the instruction manual which comes with the slide rule, together with some supervised instruction (if possible) in the classroom or laboratory.

Since the theory of the slide rule is based on *logarithms*, we begin our discussion with a review of the definition and properties of logarithms.

8.2 LOGARITHMS

The concept of a *logarithm* is a natural extension of the discussion of *exponents* in Chapter 1. By definition,

> The logarithm to the base a of a number x *is equal to* the exponent y to which the base number a must be raised in order that $x = a^y$. (8.1)

That is, if

$$x = a^y \qquad\qquad (8.2)$$

then

$$y = \log_a x \qquad\qquad (8.3)$$

In Equations 8.2 and 8.3, a is referred to as the *base number*. Equation 8.3 is the mathematical representation of the statement in Equation 8.1, and is to be read as

"*y* is equal to the logarithm to the base *a* of *x*"

Therefore, since $2^5 = 32$, we conclude that

$$5 = \log_2 32$$

Similarly, since $10^3 = 1000$, we find

$$3 = \log_{10} 1000$$

Furthermore, since $3^2 = 9$, it follows that

$$2 = \log_3 9$$

Since the logarithms of numbers are *exponents*, they have the same mathematical properties as exponents. For example, if

$$A = a^p, \text{ and } B = a^q$$

we find that

$$p = \log_a A, \text{ and } q = \log_a B$$

Therefore,

$$p + q = \log_a A + \log_a B$$

However, we also know (by generalizing Eq. 1.1)

$$AB = a^p a^q = a^{p+q}$$

and so we find that

$$p + q = \log_a AB$$

The above expressions for $p + q$ can be combined to give the important identity,

$$\log_a AB = \log_a A + \log_a B \qquad (8.4)$$

That is, the logarithm of the *product* of two numbers is equal to the *sum* of the logarithms of the individual numbers. The identity in Equation 8.4 forms the operational basis for multiplying two numbers, A and B, on a slide rule. That is, we *multiply* two numbers on a slide rule by *adding* their logarithms. In a similar manner (by using Eq. 1.4), we have

$$\log_a \frac{A}{B} = \log_a A - \log_a B \qquad (8.5)$$

That is, the logarithm of the *ratio* of two numbers is equal to the *difference* of the logarithms of the individual numbers. Equation 8.5 forms the operational basis for *dividing* two numbers, A by B, on a slide rule, namely, we *subtract* their logarithms. Finally, we leave it as an exercise for the student to verify the identity,

$$\log_a A^n = n \log_a A \qquad (8.6)$$

That is, the logarithm of a number to a given exponent n is equal to n times the logarithm of the number.

From Equation 8.4 we conclude that

$$\log_a (3 \cdot 2) = \log_a 3 + \log_a 2$$

Furthermore, from Equation 8.5 we find

$$\log_a \frac{3}{2} = \log_a 3 - \log_a 2$$

In addition, from Equation 8.6 it follows that

$$\log_a 3^2 = 2 \log_a 3$$

Logarithms to the base 10 are called *common logarithms*. Rather than write $\log_{10} x$, it is customary to use the short-form notation,

$$\log x \qquad \text{(common logarithm)}$$

to denote the common logarithm of x. A table of values of $\log x$ is given on page 214 of this book for values of x ranging from 1.0 to 9.9, in steps of 0.1. For our present purposes we have tabulated in Table 8.1 the values of $\log x$, for x ranging from 1 to 10, in steps of unity. Notice that

TABLE 8.1 LOGARITHMS TO THE BASE 10

x	$\log x$
1	0.000
2	0.301
3	0.477
4	0.602
5	0.699
6	0.778
7	0.845
8	0.903
9	0.954
10	1.000

the first and last values of log x in Table 8.1 follow from $10^0 = 1$ and $10^1 = 10$, respectively. We conclude this section with some simple numerical examples which make use of Table 8.1 and Equations 8.4 and 8.5. These examples will be useful in subsequent sections to understand the basic operations and scale sizes associated with the slide rule.

Example 8.2.1

$$\log 10 - \log 1 = \,?$$

From Equation 8.5 and Table 8.1, we find

$$\log 10 - \log 1 = \log \frac{10}{1} = 1.000$$

Example 8.2.2

$$\log 2 - \log 1 = \,?$$

From Equation 8.5 and Table 8.1, we find

$$\log 2 - \log 1 = \log \frac{2}{1} = 0.301$$

Example 8.2.3

$$\log 2 + \log 3 = ?$$

From Equation 8.4 and Table 8.1, we find

$$\log 2 + \log 3 = \log(2 \cdot 3) = \log 6 = 0.778$$

Also, by direct computation,

$$\log 2 + \log 3 = 0.301 + 0.477 = 0.778$$

Example 8.2.4

$$\log 3 - \log 2 = ?$$

From Table 8.1, we have

$$\log 3 - \log 2 = 0.477 - 0.301 = 0.176$$

Also, from Equation 8.5, we find

$$\log 3 - \log 2 = \log \frac{3}{2} = \log 1.5$$

Therefore,

$$\log 1.5 = 0.176$$

EXERCISES

1. Verify the identity in Equation 8.5.
2. Verify the identity in Equation 8.6.
3. $\log 4 - \log 2 =$ ___ (Ans. 156)
4. $\log 4 + \log 2 =$ ___ (Ans. 85)
5. $\log 9 - \log 1 =$ ___ (Ans. 338)
6. $\log 9 + \log 1 =$ ___ (Ans. 237)
7. $\log 16 = 4 \log$ ___ (Ans. 280)

8. log 16 = ____   (Ans. 236)

9. log 8 = 3 log ____   (Ans. 36)

10. log 5 − log 2 = ____   (Ans. 297)

8.3 THE PRINCIPLE OF SLIDE-RULE OPERATION

As illustrated in Figure 8.1, the C scale of a standard 10-inch slide rule is located on the moveable *slide*, whereas the D scale is located on the stationary frame of the slide rule. The spacings between the numbers on the two scales are identical.

On both the C and D scales it is customary to refer to the left-most division (labeled by the number 1) as the *left index*, and the right-most division (also labeled by the number 1) as the *right index*. Note from Figure 8.1 that the C and D scales are *not* uniform. In fact, the positions of the numbers on the C and D scales are determined from the values of their logarithms. For example, since

$$\log 2 - \log 1 = 0.301$$

(see Table 8.1), the number 2 is located (on a 10-inch slide rule)

$$0.301 \times 10 \text{ in.} = 3.01 \text{ in.}$$

to the right of the left index. Furthermore, since

$$\log 4 - \log 1 = 0.602,$$

the number 4 is located

$$0.602 \times 10 \text{ in.} = 6.02 \text{ in.}$$

to the right of the left index. The locations of the numbers 3, 5, 6, etc., on the C and D scales can be determined in a similar manner. (Measure these positions and verify these calculations.)

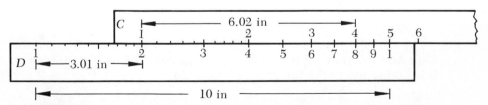

FIGURE 8.1   The C and D scales of a slide rule (the positions of the numbers are determined by the values of their common logarithms).

The underlying principle by which we *multiply* or *divide* on a slide rule is based on the *addition* or *subtraction* of logarithms (Eqs. 8.4 and 8.5).

Example 8.3.1

As a simple example which illustrates the principle underlying *multiplication* on a slide rule, we consider the slide-rule setting shown in Figure 8.1. To determine the value of the product 2×4, the left index of the C scale is placed directly *above* the number 2 on the D scale. Reading across the C scale to the number 4, we find that 8 ($=2 \times 4$) is located directly *below* on the D scale. Evidently, the net result of this operation is to verify that $2 \times 4 = 8$. It is important to realize that what has really occurred in this *multiplication* process is the *addition* of two distances proportional to log 2 and log 4. Referring to Figure 8.1 and Table 8.1, we note that

3.01 in. is proportional to log 2 (i.e., 0.301)

6.02 in. is proportional to log 4 (i.e., 0.602)

Adding, we find 3.01 in. + 6.02 in. = 9.03 in., which is proportional to (see Eq. 8.4)

$$\log 2 + \log 4 = \log(2 \times 4) = \log 8 \text{ (i.e., 0.903)}$$

That is,

9.03 in. is proportional to log 8.

Therefore, the two numbers 2 and 4 have been multiplied by adding distances proportional to their logarithms.

Example 8.3.1 illustrates the general method for multiplication on a slide rule. For any two numbers, A and B, the product AB is obtained by adding two distances proportional to log A and log B, and making use of the identity (see Eq. 8.4),

$$\log A + \log B = \log AB$$

In a similar manner, the division of two numbers, A by B, is accomplished by subtracting two distances proportional to log A and log B, and making use of the identity (see Eq. 8.5),

$$\log A - \log B = \log \frac{A}{B}$$

Example 8.3.2

As a simple example which illustrates the principle underlying *division* on a slide-rule, we again consider the slide-rule setting shown in Figure 8.1. To determine the value of the ratio 8/4 the number 4 on the C scale is placed directly above 8 on the D scale. Reading across the C scale to the left index (1), we find that the number 2 (=8/4) is located directly below on the D scale. The net result of this operation is to verify that 8/4 = 2. What has occurred in this division process is the *subtraction* of two distances proportional to log 8 and log 4. Referring to Figure 8.1 and Table 8.1 we note that

9.03 in. is proportional to log 8 (i.e., 0.903)

6.02 in. is proportional to log 4 (i.e., 0.602)

Subtracting, we find 9.03 in. − 6.02 in. = 3.01 in., which is proportional to (see Eq. 8.5)

$$\log 8 - \log 4 = \log \frac{8}{4} = \log 2 \text{ (i.e., 0.301)}$$

That is,

3.01 in. is proportional to log 2

Therefore, the value of 8 divided by 4 has been determined by subtracting distances proportional to their logarithms.

EXERCISES

1. For the slide-rule setting shown in Figure 8.1, verify that

    (a) 2 (*D* scale) × 2 (*C* scale) = 4 (*D* scale)

    (b) 2 (*D* scale) × 3 (*C* scale) = 6 (*D* scale)

    (c) 2 (*D* scale) × 5 (*C* scale) = 10 (*D* scale)

    (d) 2 (*D* scale) × 1.5 (*C* scale) = 3 (*D* scale)

[Hint: Follow the procedure outlined in Example 8.3.1. Note that in each case the left index (1) on the *C* scale is located directly above the number 2 on the *D* scale (2 is the first factor in each of the above products).]

2. For the slide-rule setting shown in Figure 8.1, verify that

    (a) 10 (*D* scale) divided by 5 (*C* scale) = 2 (*D* scale)

(b) 6 (D scale) divided by 3 (C scale) = 2 (D scale)

(c) 4 (D scale) divided by 2 (C scale) = 2 (D scale)

(d) 3 (D scale) divided by 1.5 (C scale) = 2 (D scale)

[Hint: Follow the procedure outlined in Example 8.3.2. In each case, note that the divisor (on the C scale) is located directly above the dividend (on the D scale). The answer 2 (for each exercise) is located on the D scale, directly below the left index (1) of the C scale.]

8.4 LOCATING NUMBERS ON THE SLIDE RULE

In this section we illustrate the method of locating various three-digit numbers on the D scale.* The student should first carefully examine the D scale and make the following observations:

(a) The D scale is divided into *ten* major divisions labeled by the large digits 1, 2, 3, 4, · · ·. These divisions represent the integers 1 through 10.

(b) The portion of the D scale between 1 and 2 is divided into *ten* minor divisions, each labeled by the small integers 1, 2, 3, 4, · · ·. These divisions correspond to the numbers 1.1, 1.2, 1.3, 1.4, · · ·. Furthermore, between each minor division there are ten smaller subdivisions. These subdivisions represent the numbers 1.11, 1.12, 1.13, 1.14, · · ·, and so on.

(c) The nine remaining portions of the D scale (between the major divisions, 2 and 3, 3 and 4, 4 and 5, · · ·, etc.) are also each divided into ten minor divisions. These minor divisions represent the numbers 2.1, 2.2, 2.3, · · ·, 3.1, 3.2, 3.3, · · ·, 4.1, 4.2, 4.3, · · ·, and so on. Note that these minor divisions are *not* labeled by small integers 1, 2, 3, 4, · · · (in contrast to that portion of the D scale between 1 and 2).

(d) The ten minor divisions between the numbers 2 and 3, and 3 and 4, are each divided into *five* smaller subdivisions representing the numbers 2.02, 2.04, 2.06, · · ·, and 3.02, 3.04, 3.06, · · ·.

(e) The ten minor divisions between the numbers 4 and 5, 5 and 6, 6 and 7, · · ·, are each divided into *two* smaller subdivisions representing the numbers 4.05, 4.10, 4.15, · · ·, 5.05, 5.10, 5.15, · · ·, and so on.

We now illustrate how to locate numbers on a slide rule with some specific examples.

---

*Numbers on the C scale can be located in an analogous manner, since the spacings between numbers on the C and D scales are identical.

Example 8.4.1

Locate the number 1.42 on the D (or C) scale.

The solution to this example is represented by the dashed line in the accompanying diagram.

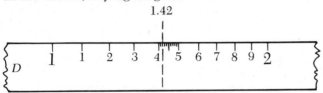

The steps used in locating the number 1.42 can be summarized as follows:

Step 1. 1.4 lies between the major divisions representing 1 and 2.

Step 2. 1.42 lies between the minor divisions representing 1.4 and 1.5.

Step 3. There are *ten* subdivisions between 1.4 and 1.5, and hence each subdivision is 0.01. Therefore, the number 1.42 is located two subdivisions to the right of 1.4, as indicated by the dashed line in the above diagram.

Example 8.4.2

Locate the number 3.77 on the D (or C) scale.

The solution to this example is represented by the dashed line in the accompanying diagram.

The steps used in locating the number 3.77 can be summarized as follows:

Step 1. 3.7 lies between the major divisions representing 3 and 4.

Step 2. 3.77 lies between the minor divisions representing 3.7 and 3.8.

Step 3. There are *five* subdivisions between 3.7 and 3.8, and hence each subdivision is 0.02. Therefore, 3.77 is located approximately one-half of the distance between the third and fourth subdivisions, as indicated by the dashed line in the above diagram.

Example 8.4.3

Locate the number 7.65 on the D (or C) scale.

The solution to this example is represented by the dashed line in the accompanying diagram.

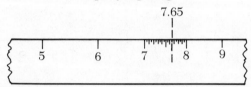

The steps used in locating the number 7.65 can be summarized as follows:

Step 1. 7.6 lies between the major divisions representing 7 and 8.

Step 2. 7.65 lies between the minor divisions representing 7.6 and 7.7.

Step 3. There are *two* subdivisions between 7.6 and 7.7, and hence each subdivision is 0.05. Therefore, 7.65 is located one subdivision to the right of 7.6, as indicated by the dashed line in the above diagram.

The discussion thus far has been limited to numbers ranging between 1 and 10. However, the D (or C) scales can, in fact, be used to represent *any* number. The only difference in locating the number is that *the decimal point must be specified independently of the slide-rule operation.*

Therefore, the setting illustrated in Example 8.4.1 can equally well represent the numbers 0.142, 0.0142, 14.2, 142, 1.42×10^5, 1.42×10^{90}, etc.

Similarly, the setting illustrated in Example 8.4.2 can also represent the numbers 377, 37.7, 3.77×10^{16}, 0.0377, 0.00377, 3.77×10^{-8}, etc.

The essential point to keep in mind is that locating a number on the D (or C) scale only determines the *significant figures*. The location of the decimal point must be specified independently.

EXERCISES

Locate the following numbers on the D and C scales.

1. 1.18	6. 7.26
2. 3.90	7. 2.54
3. 4.55	8. 5.14
4. 8.90	9. 6.83
5. 9.75	10. 9.10

8.5 MULTIPLICATION AND DIVISION ON THE SLIDE RULE

To multiply or divide on the slide rule we make use of the C and D scales in the manner indicated in Examples 8.3.1 and 8.3.2. In general, to determine the value of the product of any two numbers, x and y, we use the procedure summarized below.

> *Multiplication on the Slide Rule*
> To evaluate the product $x \cdot y$:
> **Step 1.** Set the left index (or right index, when appropriate) of the C scale directly above the first number, x, on the D scale.
> **Step 2.** Adjust the position of the moveable *cursor* so that its *hairline* is located on the second number, y, on the C scale.
> **Step 3.** The value of the product, xy, is located directly below the hairline on the D scale.

Example 8.5.1

Evaluate 1.55×2.58 on the slide rule.

The solution to this problem is illustrated in the accompanying figure.

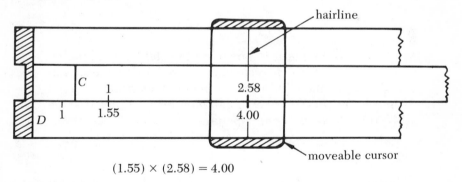

$$(1.55) \times (2.58) = 4.00$$

In this example, the *left index* of the C scale is located directly above the number 1.55 on the D scale. The position of the cursor is adjusted so that the hairline is on the number 2.58 on the C scale. The value of the product (to three significant figures),

$$4.00 = 1.55 \times 2.58$$

is located directly below the hairline on the D scale.

Example 8.5.2

Evaluate 5.50×4.55 on the slide rule.
The solution to this problem is illustrated in the accompanying figure.

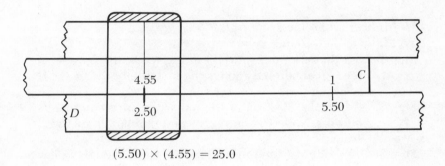

$$(5.50) \times (4.55) = 25.0$$

In this example the *right index* of the C scale is located directly above the number 5.50 on the D scale. The position of the cursor is adjusted so that the hairline is on the number 4.55 on the C scale. The value of the product (to three significant figures) is located directly below the hairline on the D scale. We find

$$25.0 = 5.50 \times 4.55$$

For this example it is important to note that the number which we read off the D scale (2.50) includes only the *significant figures* in the final answer. *The location of the decimal point in the final answer (25.0) has been specified independently of the slide-rule operation.*

It is important to note that in Example 8.5.1 the *left index* of the C scale was placed above the first factor (1.55) in the product, whereas in Example 8.5.2 the *right index* of the C scale was placed above the first factor (5.50) in the product. As a general rule, we use whichever index (left *or* right) is necessary, so that the *second factor* in the product (that is, 2.58 in Example 8.5.1, or 4.55 in Example 8.5.2) does *not* extend over the end of the fixed scale. For example, Figure 8.2 illustrates the *incorrect* method for attempting to evaluate the product 5.50×4.55. We note here that the second factor (4.55) in the product extends over the end of the fixed scale, and the multiplication process cannot be carried out. The correct method for evaluating the product 5.50×4.55 is illustrated in the figure accompanying Example 8.5.2. That is, the *right index* (rather

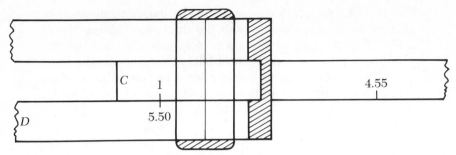

FIGURE 8.2 *Incorrect* method for attempting to evaluate the product 5.50 × 4.55. The *correct* method is illustrated in the figure in Example 8.5.2 (where the *right index* of the C scale is placed over the factor 5.50).

than the *left index*) of the C scale is located above the factor 5.50 in this case.

In general, to evaluate the quotient of any two numbers, x divided by y, we use the procedure summarized below.

Division on the Slide Rule

To evaluate the quotient $\frac{x}{y}$:

Step 1. Adjust the position of the cursor so that its hairline is directly over the numerator, x, on the D scale.

Step 2. Move the slide so that the denominator, y, falls directly below the hairline of the cursor on the C scale.

Step 3. The value of the quotient, $\frac{x}{y}$, is located on the D scale, directly below the index (left or right, where appropriate) of the C scale.

Example 8.5.3

Evaluate 3.50/2.16 on the slide rule.

The solution to this problem is illustrated in the accompanying figure.

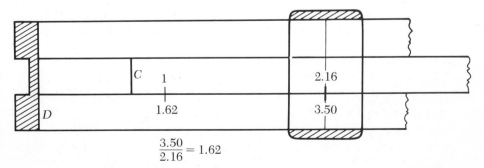

$$\frac{3.50}{2.16} = 1.62$$

In this example the position of the cursor is adjusted so that its hairline is directly over the numerator 3.50 on the D scale. The position of the slide is adjusted so that the denominator 2.16 (on the C scale) falls directly under the hairline. The quotient (to three significant figures) is located on the D scale, directly below the *left index* of the C scale. That is,

$$\frac{3.50}{2.16} = 1.62$$

Note that the slide-rule setting for this example is identical to the setting required for evaluating the product

$$1.62 \times 2.16 = 3.50$$

Example 8.5.4

Evaluate 2.40/4.95 on the slide rule.

The solution to this problem is illustrated in the accompanying figure.

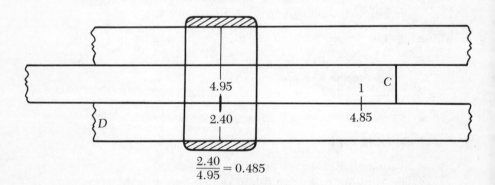

$$\frac{2.40}{4.95} = 0.485$$

In this example the position of the cursor is adjusted so that its hairline is directly over the numerator 2.40 on the D scale. The position of the slide is adjusted so that the denominator 4.95 (on the C scale) falls directly under the hairline. The value of the quotient (to three significant figures) is located on the D scale directly below the *right index* of the C scale. We find

$$\frac{2.40}{4.95} = 0.485$$

For this example, the number which we read off the D scale (4.85) includes only the *significant figures* in the final answer. The location of the decimal point in the final answer (0.485) has been specified independently of the slide-rule operation.

Note that the slide-rule setting for this example is identical to the setting required for evaluating the product

$$0.485 \times 4.95 = 2.40$$

Many calculations require a series of successive multiplications and divisions. The general procedure for this type of problem may be summarized as follows:

(1) Write all the numbers in powers-of-ten notation.
(2) Make an *order-of-magnitude estimate* of the answer to determine the location of the decimal point.
(3) Carry out the multiplication and division process on the slide rule to find the significant figures in the final answer.

Example 8.5.5

Evaluate $\dfrac{(275)(0.125 \times 10^{-5})}{(11.3)(3.85)}$ on the slide rule.

We first write all factors in powers-of-ten notation; this gives

$$\frac{(2.75 \times 10^2)(1.25 \times 10^{-6})}{(1.13 \times 10)(3.85)},$$

which reduces to

$$\frac{2.75 \times 1.25 \times 10^{-4}}{1.13 \times 3.85 \times 10} = \frac{2.75 \times 1.25}{1.13 \times 3.85} \times 10^{-5}$$

To determine the location of the decimal point in the final answer we make an *order-of-magnitude estimate*:

$$\frac{2.75 \times 1.25}{1.13 \times 3.85} \times 10^{-5} \approx \frac{3 \times 1}{1 \times 4} \times 10^{-5} = 0.75 \times 10^{-5}$$

Having established the powers-of-ten and location of the decimal point in the final answer, we now proceed to evaluate the combination of factors

$$\frac{2.75 \times 1.25}{1.13 \times 3.85}$$

To do this we first divide 2.75 by 1.13, then multiply by 1.25, and finally divide by 3.85. That is, we *multiply and divide, alternating the processes.* In practice, this saves considerable time in the calculation. The steps are summarized as follows:

Step 1. In order to divide 2.75 by 1.13, we adjust the position of the cursor so that its hairline is directly over 2.75 on the D scale. The position of the slide is then adjusted so that 1.13 (on the C scale) is directly under the hairline. The quotient (2.43), *which need not be written down,* is then located on the D scale directly below the left index of the C scale.

Step 2. To multiply the result of Step 1 by the factor 1.25, we adjust the position of the cursor so that its hairline is directly above 1.25 on the C scale. The value of the product (3.04), *which need not be recorded,* is then located under the hairline on the D scale. The slide rule is now in a position to carry out the division by the factor 3.85.

Step 3. To divide by 3.85, we leave the cursor where it is, and move the slide until 3.85 on the C scale is directly under the hairline. We then read the significant figures of the final answer, 7.90, on the D scale directly below the right index of the C scale. Comparing with the order-of-magnitude estimate, we conclude that

$$\frac{2.75 \times 1.25}{1.13 \times 3.85} \times 10^{-5} = 0.790 \times 10^{-5} = 7.90 \times 10^{-6}$$

EXERCISES

Use the slide rule to evaluate the following:

1. $7.50 \times 3.62 =$ ____   (Ans. 206)

2. $1.43 \times 6.95 =$ ____   (Ans. 353)

3. $18.7 \times 1.25 =$ ____   (Ans.   38)

4. $(9.20 \times 10^{-5}) \times (1.63 \times 10^{2}) =$ ____   (Ans.   79)

5. $\dfrac{7.85}{3.20} =$ ____   (Ans. 279)

6. $\dfrac{5.67}{18.9} =$ ____   (Ans. 204)

7. $\dfrac{2.63 \times 10^{-5}}{3.96 \times 10^{10}} = $ ____ (Ans. 196)

8. $\dfrac{1.65 \times 3.40}{8.90 \times 7.35} = $ ____ (Ans. 138)

9. $\dfrac{(5.50 \times 10^{11}) \times (2.75 \times 10^{-6})}{(3.33 \times 10^{-5}) \times (1.65 \times 10^{8})} = $ ____ (Ans. 239)

10. $\dfrac{(3.14 \times 1.05 \times 9.80)}{(1.55 \times 2.65 \times 7.55)} = $ ____ (Ans. 306)

8.6 SQUARES AND SQUARE ROOTS ON THE SLIDE RULE

The A and D scales are used to square a number, or take its square root. In fact, the slide can be completely removed from the rule for these operations.* As illustrated in Figure 8.3, the A scale consists of *two* ranges, each running from 1 to 1.

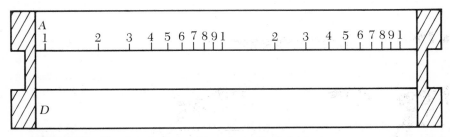

FIGURE 8.3 The A scale on a slide rule is divided into two ranges, each running from 1 to 1.

In practice, it is convenient to think of the index 1 at the center of the A *scale as representing the number 10, and the index 1 at the extreme right of the* A *scale as representing the number 100.* In this case we may view the left half of the A scale as representing the range of numbers 1 through 10, and the right half of the A scale as representing the range of numbers 10 through 100.

In general, to square a number x on the slide rule, we use the procedure summarized below.

Taking Squares on the Slide Rule
To evaluate x^2:
Step 1. Adjust the position of the cursor so that its hairline is directly over the number x on the D scale.
Step 2. The value of the square, x^2, is located directly under the hairline on the A scale.

*For purposes of learning how to take squares and square roots, the student may find it less confusing if the slide is completely removed.

Using this procedure, it is readily verified by inspection of the A and D scales that $2^2 = 4$, $3^2 = 9$, $4^2 = 16$, $5^2 = 25$, etc.

Example 8.6.1

Evaluate $(2.62 \times 10^8)^2$ using the slide rule.
Using the rules of exponents we first note that

$$(2.62 \times 10^8)^2 = (2.62)^2 \times 10^{16}$$

The slide-rule setting which is required to evaluate $(2.62)^2$ is illustrated in the accompanying figure.

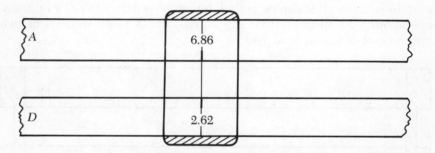

In this example the position of the cursor is adjusted so that its hairline is directly over 2.62 on the D scale. The value of the square is located directly below the hairline on the A scale. That is,

$$(2.62)^2 = 6.86$$

Therefore, we conclude that

$$(2.62 \times 10^8)^2 = 6.86 \times 10^{16}$$

[Note: For this example there is actually some difficulty in reading the answer to three-figure accuracy. In particular, due to the rather coarse resolution on the A scale, it is difficult to distinguish 6.86 from 6.85.]

To extract the square root of a number on the slide rule, the procedure is essentially the reverse of that outlined above.

Extracting Square Roots on the Slide Rule

If the number whose square root we are extracting falls between 1 and 100:*

Step 1. Adjust the position of the cursor so that its hairline is over the number on the *A* scale.

Step 2. The square root is located under the hairline on the *D* scale.

If the number whose square root we are extracting does *not* lie between 1 and 100, then we rewrite the number in powers of ten so that the coefficient is a number between 1 and 100, and the exponent of ten is an *even* power (that is, 0, 2, 4, $\cdots$, -2, -4, $\cdots$). We then extract the square root following the procedure outlined above.

Using this procedure it is readily verified that $(81)^{\frac{1}{2}} = 9$, $(64)^{\frac{1}{2}} = 8$, $(49)^{\frac{1}{2}} = 7$, $(4)^{\frac{1}{2}} = 2$, etc.

As a less trivial example, we evaluate the square root

$$(62.5 \times 10^{11})^{\frac{1}{2}}$$

Following the procedure outlined above, we first write 62.5×10^{11} in the form

$$62.5 \times 10^{11} = 6.25 \times 10^{12}$$

Note that 6.25 lies between 1 and 100, and that 10^{12} is an even power of 10. We thus find

$$(62.5 \times 10^{11})^{\frac{1}{2}} = (6.25 \times 10^{12})^{\frac{1}{2}} = (6.25)^{\frac{1}{2}} \times (10^{12})^{\frac{1}{2}}$$

To determine $(6.25)^{\frac{1}{2}}$ we adjust the position of the cursor so that its hairline is directly over 6.25 on the *A* scale. The square root of 6.25 is then located under the hairline on the *D* scale. That is, $(6.25)^{\frac{1}{2}} = 2.50$. Therefore, we conclude that

$$(62.5 \times 10^{11})^{\frac{1}{2}} = 2.50 \times 10^{6}$$

---

*Keep in mind that the left half of the *A* scale corresponds to the range 1 through 10, and the right half of the *A* scale corresponds to the range 10 through 100.

EXERCISES

Use the slide rule to evaluate:

1. $(3.10)^{\frac{1}{2}} = $ ____   (Ans. 300)

2. $(13.5)^{\frac{1}{2}} = $ ____   (Ans. 255)

3. $(2.50)^{\frac{1}{2}} = $ ____   (Ans. 224)

4. $(76.5)^{\frac{1}{2}} = $ ____   (Ans. 113)

5. $(1.65 \times 10^{-5})^{\frac{1}{2}} = $ ____   (Ans. 68)

6. $(96.1 \times 10^{81})^{\frac{1}{2}} = $ ____   (Ans. 67)

7. $(150)^2 = $ ____   (Ans. 163)

8. $(1.47)^2 = $ ____   (Ans. 337)

9. $(23.4)^2 = $ ____   (Ans. 273)

10. $(4.55 \times 10^6)^2 = $ ____   (Ans. 261)

PHYSICAL UNITS AND CONVERSIONS

In this chapter we review the situation with regard to physical units and the methods for conversion from one system of units to another. We will first give a summary of *mechanical* units, repeating some of the discussion presented in Chapter 1. Then, we will take up the question of *electrical* units. Finally, we will give some of the non-standard units that find application in particular situations.

9.1 BASIC UNITS OF MEASURE

The basic units of physical measure are those of *length, mass,* and *time.* All physical quantities can be expressed in terms of units that are combinations of these three basic units. (In the MKS system of units for electrical measure we introduce a fourth basic unit, that of electrical *charge.* This is done to achieve a certain degree of simplicity, but it is not necessary; in the CGS system, the unit of charge is expressed in terms of length, mass, and time.)

The systems of units currently in use are the *metric system* (including the MKS or meter-kilogram-second and the CGS or centimeter-gram-second variations) and the *British engineering system.* Because the latter is slowly being phased out and will eventually be replaced by the metric system, we will discuss only a few of the more important and common British units of measure. The British system of units is no longer an independent system; all the fundamental British units of measure are now defined in terms of metric units. Table 9.1 summarizes the basic units of all three systems, and Tables 9.2 and 9.3 give the various conversion factors between the systems.

TABLE 9.1   BASIC UNITS OF PHYSICAL MEASURE

	MKS	CGS	British
Length	meter (m)	centimeter (cm)	yard (yd)
Mass	kilogram (kg)	gram (g)	pound-mass (lb)
Time	second (s)	second (s)	second (s)

TABLE 9.2   CONVERSION FACTORS FOR THE BASIC UNITS OF LENGTH

	m	cm	yd	in
1 m =	1	100	1.094	39.37
1 cm =	0.01	1	0.01094	0.3937
1 yd =	0.9144	91.44	1	36
1 in =	0.0254	2.54	1/36	1

1 in = 2.54 cm (exactly)

1 mi = 1760 yd = 5280 ft (statute mile)

$\qquad$ = 1.609 km = 1.609×10^3 m = 1.609×10^5 cm

1 km = 0.6214 mi

Areas and volumes:

1 in^2 = 6.452 cm^2

1 ft^2 = 929 cm^2

1 in^3 = 16.39 cm^3

1 ft^3 = 2.832×10^4 cm^3

TABLE 9.3   CONVERSION FACTORS FOR THE BASIC UNITS OF MASS

	kg	g	lb
1 kg =	1	10^3	2.205
1 g =	10^{-3}	1	2.205×10^{-3}
1 lb =	0.4536	453.6	1

1 lb = 453.59237 g (exactly)

Example 9.1.1

One *acre* consists of 43,560 ft². Express this figure in m².

$$1 \text{ acre} = (4.356 \times 10^4 \text{ ft}^2) \times \left(\frac{929 \text{ cm}^2}{1 \text{ ft}^2}\right) \times \left(\frac{1 \text{ m}^2}{10^4 \text{ cm}^2}\right)$$

$$= 4046.7 \text{ m}^2$$

Example 9.1.2

The *nautical mile* was originally defined to be the distance corresponding to 1 minute of arc (1/60 of a degree) of longitude at the Earth's equator. Since 1959 the nautical mile has been defined to be 1852 m. Use this information to compute the radius of the Earth in km.

The circumference of the Earth in nautical miles is

$$C = (360 \text{ deg}) \times \left(\frac{60 \text{ arc min}}{1 \text{ deg}}\right) \times \left(\frac{1 \text{ naut mi}}{1 \text{ arc min}}\right)$$

$$= 2.16 \times 10^4 \text{ naut mi}$$

Since $C = 2\pi R$, the Earth's radius is

$$R = \frac{C}{2\pi} = \frac{1}{2\pi} \times (2.16 \times 10^4 \text{ naut mi}) \times \left(\frac{1852 \text{ m}}{1 \text{ naut mi}}\right) \times \left(\frac{1 \text{ km}}{10^3 \text{ m}}\right)$$

$$= 6.38 \times 10^3 \text{ km}$$

EXERCISES

1. Express 1 mi² in cm². (Ans. 229)

2. 1 gallon = 231 in³. Express this figure in cm³. (Ans. 254)

3. How many acres are there in 1 mi²? (Refer to Example 9.1.1). (Ans. 72)

4. One metric ton is equal to 10^3 kg. Express this figure in lb. (Ans. 50)

9.2 MECHANICAL QUANTITIES

Table 9.4 lists the various physical quantities that we encounter in discussions of mechanics. (This list is restricted to those quantities that the reader is likely to use in an introductory physics course.) The dimensions of each quantity are given in terms of the basic units of length (L), mass (M), and time (T), and the special names used for the derived units in the MKS and CGS systems are given in the last two columns. These derived units are defined in Table 9.5.

TABLE 9.4 DIMENSIONS AND UNITS OF MECHANICAL QUANTITIES

Quantity	Dimension	Metric Units	
		MKS	CGS
Acceleration, $\vec{a}$	LT^{-2}	m/s²	cm/s²
Angle, θ	——	rad	rad
Angular acceleration, $\vec{\alpha}$	T^{-2}	rad/s²	rad/s²
Angular frequency, ω	T^{-1}	rad/s	rad/s
Angular momentum, $\vec{L}$	ML^2T^{-1}	kg-m²/s	g-cm²/s
Angular velocity, $\vec{\omega}$	T^{-1}	rad/s	rad/s
Area, A	L^2	m²	cm²
Density, ρ	ML^{-3}	kg/m³	g/cm³
Displacement, $\vec{x}, \vec{s}$,	L	m	cm
Energy, E	ML^2T^{-2}	J	erg
Energy (thermal), Q	ML^2T^{-2}	Cal	cal
Force, $\vec{F}$	MLT^{-2}	N	dyne
Frequency, ν	T^{-1}	Hz (s⁻¹)	Hz (s⁻¹)
Inertia (rotational), I	ML^2	kg-m²	g-cm²
Length, l	L	m	cm
Mass, m	M	kg	g
Momentum (linear), $\vec{p}$	MLT^{-1}	kg-m/s	g-cm/s
Period, τ	T	s	s
Power, P	ML^2T^{-3}	W	erg/s
Pressure, p, P	$ML^{-1}T^{-2}$	N/m²	dyne/cm²
Time, t	T	s	s
Torque, $\vec{\tau}$	ML^2T^{-2}	N-m	dyne-cm
Velocity, $\vec{v}$	LT^{-1}	m/s	cm/s
Volume, V	L^3	m³	cm³
Wavelength, λ	L	m	cm
Work, W	ML^2T^{-2}	J	erg

TABLE 9.5 DERIVED UNITS

Name	Equivalence
calorie	1 cal = 4.186×10^7 erg = 10^{-3} Cal
Calorie	1 Cal = 10^3 cal = 4186 J
dyne	1 dyne = 1 g-cm/s² = 10^{-5} N
erg	1 erg = 1 g-cm²/s² = 1 dyne-cm = 10^{-7} J
joule	1 J = 1 kg-m²/s² = 1 N-m = 10^7 erg
newton	1 N = 1 kg-m/s² = 10^5 dyne
watt	1 W = 1 N-m/s = 1 J/s = 10^7 erg/s

Example 9.2.1

Newton's law of universal gravitational attraction states that the force between two masses, m_1 and m_2, separated by a distance r is

$$F = G \frac{m_1 m_2}{r^2} \tag{1}$$

where G is the *universal gravitational constant*. What are the dimensions of G?

Solving (1) for G, we find

$$G = \frac{Fr^2}{m_1 m_2} \tag{2}$$

and substituting the dimensions for each quantity on the right-hand side, we have

$$G = \frac{(MLT^{-2}) \times (L)^2}{(M) \times (M)} = \frac{L^3 T^{-2}}{M} \tag{3}$$

In MKS and CGS units, the value of G is

$$G = 6.673 \times 10^{-11} \text{ N-m}^2/\text{kg}^2 \text{ (MKS)} \tag{4}$$

$$= 6.673 \times 10^{-8} \text{ dyne-cm}^2/\text{g}^2 \text{ (CGS)} \tag{5}$$

Verify that the units in (4) and (5) do, in fact, have the dimensions given in (3).

Example 9.2.2

What is the kinetic energy (in J) of a 50-lb block that is moving with a velocity of 30 ft/s?

According to Table 1.4, kinetic energy is given by

$$E_K = \frac{1}{2} mv^2$$

Before we use this equation, however, we convert the mass and the velocity of the block to metric units:

$$m = 50 \text{ lb} = (50 \text{ lb}) \times \left(\frac{0.4536 \text{ kg}}{1 \text{ lb}} \right) = 22.68 \text{ kg}$$

$$v = 30 \text{ ft/s} = (30 \text{ ft/s}) \times \left(\frac{1 \text{ yd}}{3 \text{ ft}} \right) \times \left(\frac{0.9144 \text{ m}}{1 \text{ yd}} \right) = 9.144 \text{ m/s}$$

Then,

$$E_K = \frac{1}{2} \times (22.68 \text{ kg}) \times (9.144 \text{ m/s})^2$$

$$= 946 \text{ kg-m}^2/\text{s}^2 = 946 \text{ J}$$

Example 9.2.3

A force of 40 N is applied to a block and moves the block a distance of 120 m in 8 min at constant velocity. How much work was done and how much power was required?

$$W = Fs = (40 \text{ N}) \times (120 \text{ m}) = 4800 \text{ N-m} = 4800 \text{ J}$$

The power required is equal to the *rate* at which work is done; that is,

$$P = \frac{W}{t} = \frac{4800 \text{ J}}{8 \times 60 \text{ s}} = 10 \text{ J/s} = 10 \text{ W}$$

EXERCISES

1. A particle moves uniformly along a straight line and at $t = 0$ is at $x = 45$ cm; at $t = 12$ s, the particle is at $x = 225$ cm. What is the velocity of the particle? (Ans. 334)

2. How many Calories are expended in raising a 20-kg block through a height of 30 m? (Use $W = mgh$.) (Ans. 34)

3. 1 horsepower (h.p.) = 746 W. How long will it require for a ¾ h.p. engine (working at 100% efficiency) to raise a 10-kg block from the ground to the top of a 40-m building? (Ans. 159)

4. A constant force of 20 N is applied to a 1-kg object originally at rest. After 10 s the object has moved a distance of 10 m. How much power is the force developing at the end of the 10-m run? (Calculate the velocity at $t = 10$ s; show that the power can be expressed as $P = Fv$.) (Ans. 219)

5. The law of momentum conservation states that the total linear momentum of an isolated system remains constant. That is, in any physical process, $p(\text{before}) = p(\text{after})$. A 15-g bullet is fired with a velocity of 3×10^4 cm/s into a stationary 75-g block of wood and remains embedded in the wood. What is the velocity of the bullet-block combination after impact? (Ans. 89)

9.3 THE DIFFERENCE BETWEEN THE MKS AND CGS SYSTEMS

When dealing with mechanical quantities, conversions between the MKS and CGS systems are simple and straightforward because all of the quantities are related by factors of 10. This ease of manipulation is lost, however, when we turn to electrical quantities because of a basic difference in definitions. We can best describe this difference by examining Coulomb's law in the two systems.

In the CGS system, Coulomb's law for the force between two charges, q_1 and q_2, separated by a distance r is

$$F = \frac{q_1 q_2}{r^2} \quad \text{(CGS)} \tag{9.1}$$

If q_1 and q_2 have the same sign (+ and + or − and −), the force is *repulsive*; if q_1 and q_2 have opposite signs (+ and −), the force is *attractive*.

In using the CGS system, we take the attitude that force and distance are fundamental quantities and therefore that *charge* is *defined* by Equation 9.1. This is possible because Equation 9.1 is written with a proportionality factor of *unity*, unlike the gravitational force equation (Example 9.2.1) where the proportionality factor G occurs. According to Equation 9.1,

$$q_1 q_2 = Fr^2$$

or,

$$(\text{charge})^2 = (\text{force}) \times (\text{distance})^2$$

Therefore, the dimensions of charge Q are

$$Q = (\text{force})^{\frac{1}{2}} \times (\text{distance})$$

$$= (MLT^{-2})^{\frac{1}{2}} \times (L)$$

$$= M^{\frac{1}{2}} L^{\frac{3}{2}} T^{-1} \tag{9.2}$$

Thus, the CGS unit of charge is

$$Q = \text{g}^{\frac{1}{2}}\text{-cm}^{\frac{3}{2}}\text{-s}^{-1}$$

Because of the fractional exponents, this is an inconvenient unit and so we give this quantity a special name, the *statcoulomb* (statC):

$$1 \text{ statC} = 1 \text{ g}^{\frac{1}{2}}\text{-cm}^{\frac{3}{2}}\text{-s}^{-1} \quad \text{(CGS)} \tag{9.3}$$

When using Coulomb's law in the CGS form (Eq. 9.1), if we substitute for charge in units of *statC* and for distance in units of *cm*, the force will be given in units of *dynes*.

If the ends of a length of wire are attached to the terminals of a battery, we know that charge will be transported along the wire; that is, a *current* will flow. The measure of current is the *rate* at which charge is transported. Thus, current = charge/time, or

$$I = \frac{Q}{t} \tag{9.4}$$

The CGS unit of current is defined as follows: If 1 statC of charge passes a given point in the wire in 1 s, the current is 1 statampere (statA). Thus,

$$1 \text{ statA} = \frac{1 \text{ statC}}{1 \text{ s}} \quad \text{(CGS)} \tag{9.5}$$

In the MKS system, the unit of charge is not defined by Coulomb's law. Instead, the unit of current is considered to be the fundamental quantity. Current can be defined in terms of the force that exists between two parallel wires a certain distance apart that carry the same current. (The details of how such measurements are made need not concern us here.) The MKS unit of current (defined in terms of force and distance) is the *ampere* (A), and the definition of the MKS unit of charge, the coulomb (C), is as follows: If a current of 1 A flows in a wire, then a charge of 1 C will pass a given point in the wire in 1 s. Thus,

$$1 \text{ A} = \frac{1 \text{ C}}{1 \text{ s}} \quad \text{(MKS)} \tag{9.6}$$

If we use this definition of charge, we can no longer write Coulomb's law in the form of Equation 9.1; instead, we must supply a proportionality constant and write

$$F = k \frac{q_1 q_2}{r^2} \quad \text{(MKS)} \tag{9.7}$$

Since force, distance, and charge are already defined, the proportionality constant k is fixed; k has the value

$$k = 10^{-7} c^2 \text{ m-kg/C}^2$$

where c is the speed of light. (We will not elaborate on how the speed of light has appeared in this expression!) Squaring the speed of light (see Example 2.2.1), we find, to an accuracy suitable for our purposes,

$$k = 9.0 \times 10^9 \text{ N-m}^2/\text{C}^2 \tag{9.8}$$

We frequently see the MKS form of Coulomb's law written as

$$F = \frac{1}{4\pi\epsilon_0} \frac{q_1 q_2}{r^2}$$

where the constant ϵ_0 is called the *permittivity constant*. Writing $k = 1/4\pi\epsilon_0$ and using the value of k in Equation 9.8, we find $\epsilon_0 = 8.85 \times 10^{-12} \text{ C}^2/\text{N-m}^2$.

When using Coulomb's law in the MKS form (Eq. 9.7), if we substitute for charge in units of C and for distance in units of m, and if we use the value of k expressed in Equation 9.8, the force will be given in units of N.

We now have sufficient information to compute the conversion factors that relate statC and C and also statA and A. The results are (the reader should verify these figures)

$$1 \text{ C} = 3 \times 10^9 \text{ statC} \tag{9.9}$$

$$1 \text{ A} = 3 \times 10^9 \text{ statA} \tag{9.10}$$

The factor "3" in each of these expressions is not exact, but is actually equal numerically to the speed of light (and should therefore be 2.997 925); however, "3" is sufficiently accurate for our purposes.

The *potential difference* V between two points is equal to the work per unit charge required to move a charge q between the two points:

$$V = \frac{W}{q} \tag{9.11}$$

If 1 erg of work is required to move a charge of 1 statC from point A to point B, then the potential difference between A and B is 1 *statvolt* (statV). Similarly, if 1 J is required to move a charge of 1 C from A to B, then the potential difference between A and B is 1 *volt* (V).

$$1 \text{ statV} = \frac{1 \text{ erg}}{1 \text{ statC}} \quad \text{(CGS)} \tag{9.12}$$

$$1 \text{ V} = \frac{1 \text{ J}}{1 \text{ C}} \quad \text{(MKS)} \tag{9.13}$$

Since $1 \text{ J} = 10^7$ erg and $1 \text{ C} = 3 \times 10^9$ statC, we find

$$1 \text{ statV} = 300 \text{ V} \tag{9.14}$$

Example 9.3.1

What is the force between two charges, $q_1 = +10$ C and $q_2 = -8$ C, that are separated by a distance of 2 m?

According to Equation 9.7, the magnitude of the force is

$$F = k \frac{q_1 q_2}{r^2} = (9.0 \times 10^9 \text{ N-m}^2/\text{C}^2) \times \frac{(+10 \text{ C}) \times (-8 \text{ C})}{(2 \text{ m})^2}$$

$$= -1.8 \times 10^{11} \text{ N}$$

The negative sign results from the fact that q_1 and q_2 have opposite signs and indicates that the force is *attractive*.

Example 9.3.2

A current of 15 A is flowing in a wire (this is approximately the maximum current allowed for most household wiring). How many electrons pass a given point in the wire each second?

In MKS units, the charge of an electron is

$$e = 1.602 \times 10^{-19} \text{ C} \tag{1}$$

Using Equation 9.4, we can write

$$Q = It \tag{2}$$

and the charge Q is equal to the number of electrons N multiplied by the electronic charge e:

$$Q = Ne \tag{3}$$

Combining (2) and (3), we have

$$N = \frac{Q}{e} = \frac{It}{e} = \frac{(15 \text{ A}) \times (1 \text{ s})}{1.602 \times 10^{-19} \text{ C}} = 9.37 \times 10^{19}$$

EXERCISES

1. Two equally charged objects separated by a distance of 1 m are found to repel each other with a force of 3.6×10^4 N. What is the charge of each object?   (Ans. 347)

2. Two charged particles, $q_1 = +3 \times 10^{-3}$ C and $q_2 = -6 \times 10^{-4}$ C, are attracted to each other with a force of 200 N. How far apart are the particles?   (Ans. 149)

3. Compute the electrical force (in N) for $q_1 = +30$ statC, $q_2 = +90$ statC, $r = 20$ cm.   (Ans. 361)

4. What is the electron charge in statC?   (Ans. 16)

5. How much work is required to move an electron through a potential difference of 10^6 V?   (Ans. 248)

9.4 ELECTRICAL QUANTITIES

It has become popular to adopt the MKS system for all electrical measurements, and so we limit the tabulations below to this system. Table 9.6 is the electrical equivalent of Table 9.4 and lists the dimensions and MKS units of the important electrical quantities.

TABLE 9.6 DIMENSIONS AND UNITS OF ELECTRICAL QUANTITIES

Quantity	Dimension	MKS Units
Capacitance, C	$M^{-1}L^{-2}T^2Q^2$	farad (F)
Charge, q, Q	Q	coulomb (C)
Conductivity, σ	$M^{-1}L^{-3}TQ^2$	$(\Omega \cdot m)^{-1}$
Current, I	$T^{-1}Q$	ampere (A)
Electric field strength, $\vec{E}$	$MLT^{-2}Q^{-1}$	V/m
Electric potential, V	$ML^2T^{-2}Q^{-1}$	volt (V)
EMF, $\mathscr{E}$	$ML^2T^{-2}Q^{-1}$	V
Inductance, L	ML^2Q^{-2}	henry (H)
Magnetic field strength, $\vec{H}$	$ML^{-1}Q$	A-m
Magnetic flux, Φ	$ML^2T^{-1}Q^{-1}$	weber (W)
Magnetic induction, $\vec{B}$	$MT^{-1}Q^{-1}$	W/m^2
Permeability, μ	MLQ^{-2}	H/m
Permittivity, ϵ	$M^{-1}L^{-3}T^2Q^2$	F/m
Resistance, R	$ML^2T^{-1}Q^{-2}$	ohm (Ω)
Resistivity, ρ	$ML^3T^{-1}Q^{-2}$	Ω-m
Voltage, V	$ML^2T^{-2}Q^{-1}$	V

TABLE 9.7 ASTRONOMICAL DISTANCE UNITS

Unit	Distance in Meters
Light year	1 L.Y. $= 9.46 \times 10^{15}$ m
Parsec	1 pc $= 3.08 \times 10^{16}$ m
Astronomical unit	1 A.U. $= 1.50 \times 10^{11}$ m

9.5 UNITS OF CONVENIENCE

Astronomers deal with enormous distances. The *nearest* star, for example, is at a distance of approximately 4×10^{16} m. Consequently, in order to abbreviate the writing of distances, several units of distances have been invented as a matter of convenience. One such unit is the *light year* (L.Y.), the distance that light travels in 1 yr. Using this unit, the distance to the nearest star is 4.3 L.Y. Another unit for expressing stellar distances is the *parsec* (pc), which stands for *par*allax of 1 *sec*ond of arc. An observer at a distance of 1 pc from the Sun would find that the radius of the Earth's orbit around the Sun subtends an angle of 1 second of arc. And the radius of the Earth's orbit defines another unit of distance (called the *astronomical unit* or A.U.) which is used in measuring distances in the solar system. Table 9.7 summarizes these astronomical distance units.

In atomic and nuclear phenomena we encounter physical quantities that are extremely small. For example, the energy required to ionize a hydrogen atom is approximately 2.2×10^{-20} J and the mass of the hydrogen atom is 1.67×10^{-27} kg. Just as astronomers have devised new and convenient units, so have the atomic and nuclear physicists. The masses of atoms and nuclei are measured on a scale in which the mass of a carbon atom (with 6 protons and 6 neutrons in the nucleus) is exactly 12 units. These units are called *atomic mass units* (AMU).

TABLE 9.8 SOME ATOMIC MASSES

Atom (or particle)	Mass (AMU)
Electron, e	0.000 549
Neutron, n	1.008 665
Hydrogen, H^1	1.007 825
Helium, He^4	4.002 603
Carbon, C^{12}	12.000 000
Oxygen, O^{16}	15.994 915
Iron, Fe^{56}	55.934 936

One of the advantages of this scale is that the mass of an atom with A nucleons (protons and neutrons) in the nucleus is always approximately equal to A AMU. Some atomic masses, measured in AMU, are given in Table 9.8, where

$$1 \text{ AMU} = 1.6605 \times 10^{-27} \text{ kg} \tag{9.15}$$

Atomic and nuclear energies are usually measured in terms of a unit called the *electron volt* (eV). In spite of its name, this unit is a measure of *energy*, not of voltage. The electron volt is defined as follows: If an object that carries a charge of $\pm e$ (the electronic charge) starts from rest and is accelerated through a potential difference of 1 volt, it will acquire in the process a kinetic energy of 1 eV:

$$1 \text{ eV} = 1.602 \times 10^{-19} \text{ J} \tag{9.16}$$

In these units, the energy required to ionize a hydrogen atom is 13.6 eV, and the energies of light photons in the visible region of the electromagnetic spectrum range from about 3.1 eV to 1.6 eV.

Units larger than the electron volt are convenient for many problems, especially in nuclear physics in which typical energies are in the range from $\sim 10^4$ eV to $\sim 10^7$ eV. The most frequently used units are

$$1 \text{ kiloelectron volt (keV)} = 10^3 \text{ eV} = 1.602 \times 10^{-16} \text{ J} \tag{9.17a}$$

$$1 \text{ megaelectron volt (MeV)} = 10^6 \text{ eV} = 1.602 \times 10^{-13} \text{ J} \tag{9.17b}$$

Example 9.5.1

What is the energy of a yellow light photon ($\nu = 5.1 \times 10^{14}$ Hz)?

The energy of a photon is given by

$$E = h\nu \tag{1}$$

where h is *Planck's constant:*

$$h = 6.625 \times 10^{-34} \text{ J-s} \tag{2}$$

Therefore, since 1 Hz $= 1 \text{ s}^{-1}$,

$$E = (6.625 \times 10^{-34} \text{ J-s}) \times (5.1 \times 10^{14} \text{ s}^{-1}) \times \left(\frac{1 \text{ eV}}{1.602 \times 10^{-19} \text{ J}}\right)$$

$$= 2.11 \text{ eV}$$

Example 9.5.2

A proton (nucleus of a hydrogen atom) is accelerated through a potential difference of 10^6 V, starting from rest. What is the final kinetic energy and the final velocity?

For the kinetic energy we have, simply,

$$E_K = 10^6 \text{ eV} = 1 \text{ MeV} = 1.602 \times 10^{-13} \text{ J}$$

In order to compute the final velocity, we use $E_K = \frac{1}{2} m_p v^2$, from which

$$v = \sqrt{\frac{2E_K}{m_p}}$$

$$= \sqrt{\frac{2 \times (1.60 \times 10^{-13} \text{ J})}{1.67 \times 10^{-27} \text{ kg}}}$$

so that

$$v = 1.38 \times 10^7 \text{ m/s}$$

If we had considered an electron, instead of a proton, falling through a potential difference of 10^6 V, we could not have computed the final velocity in such a simple way. Relativity theory provides the correct method of calculation. The relativistic effect, which is manifest for a 1-MeV electron, is negligible for a proton of the same energy owing to the much larger mass of the proton. However, for protons with energies of about 100 MeV or more we must also use the relativistic expression for computing the velocity.

Masses (as well as energies) can be stated in terms of eV or MeV if we make use of the result of relativity theory that expresses the correspondence between mass and energy. This is the famous Einstein mass-energy relation:

$$E = mc^2 \tag{9.18}$$

In order to compute the energy that corresponds to a mass of 1 AMU, we use Equation 9.15,

$$1 \text{ AMU} = 1.6605 \times 10^{-27} \text{ kg}$$

and multiply both sides by c^2:

$$(1 \text{ AMU}) \times c^2 = (1.6605 \times 10^{-27} \text{ kg}) \times (3 \times 10^8 \text{ m/s})^2$$

$$= (1.4945 \times 10^{-10} \text{ J}) \times \left(\frac{1 \text{ MeV}}{1.602 \times 10^{-13} \text{ J}}\right)$$

$$= 931.5 \text{ MeV}$$

so that

$$1 \text{ AMU} = 931.5 \text{ MeV}/c^2 \qquad (9.19a)$$

or

$$c^2 = 931.5 \text{ MeV/AMU} \qquad (9.19b)$$

Example 9.5.3

What is the mass-energy of a proton?
The mass of a proton is

$$m_p = 1.6726 \times 10^{-27} \text{ kg}$$

or, in terms of AMU,

$$m_p = (1.6726 \times 10^{-27} \text{ kg}) \times \left(\frac{1 \text{ AMU}}{1.6605 \times 10^{-27} \text{ kg}}\right)$$

$$= 1.0073 \text{ AMU}^*$$

Therefore, using Equation 9.19b

$$m_p c^2 = (1.0073 \text{ AMU}) \times \left(\frac{931.5 \text{ MeV}}{1 \text{ AMU}}\right)$$

$$= 938.3 \text{ MeV}$$

Some important mass-energies are given in Table 9.9.

TABLE 9.9 MASS-ENERGIES OF SOME PARTICLES

Particle	Mass (kg)	Mass-Energy (MeV)
Electron	9.11×10^{-31}	0.511
Proton	1.6726×10^{-27}	938.26
Neutron	1.6749×10^{-27}	939.55

EXERCISES

1. Express 1 A.U. in *light minutes*. (Ans. 114)

2. What is the distance to the nearest star in pc? (Ans. 291)

3. What is the mass (in kg) of an atom of O^{16}? (Ans. 245)

4. Verify Equation 9.16.

5. What is the velocity of a 1-keV electron? (Ans. 304)

6. Show that $m_e c^2 = 511$ keV. ($m_e =$ electron mass).

---

*To seven significant figures the mass of a proton is $m_p = 1.007\ 276$ AMU. Notice that m_p is just the *difference* between the mass of the hydrogen *atom* and the mass of the electron. (See Table 9.8.)

APPENDIX A.   SOME USEFUL TABLES

TABLE I. TRIGONOMETRIC TABLES OF
SINES, COSINES, AND TANGENTS
(From 0° to 90° in steps of 0.1°)

θ, Degrees	θ, Radians	Sin θ	Cos θ	Tan θ
0.0	0.0	0.0	1.0000	0.0
0.1	0.0017	0.0017	1.0000	0.0017
0.2	0.0035	0.0035	1.0000	0.0035
0.3	0.0052	0.0052	1.0000	0.0052
0.4	0.0070	0.0070	1.0000	0.0070
0.5	0.0087	0.0087	1.0000	0.0087
0.6	0.0105	0.0105	0.9999	0.0105
0.7	0.0122	0.0122	0.9999	0.0122
0.8	0.0140	0.0140	0.9999	0.0140
0.9	0.0157	0.0157	0.9999	0.0157
1.0	0.0175	0.0175	0.9998	0.0175
1.1	0.0192	0.0192	0.9998	0.0192
1.2	0.0209	0.0209	0.9998	0.0209
1.3	0.0227	0.0227	0.9997	0.0227
1.4	0.0244	0.0244	0.9997	0.0244
1.5	0.0262	0.0262	0.9997	0.0262
1.6	0.0279	0.0279	0.9996	0.0279
1.7	0.0297	0.0297	0.9996	0.0297
1.8	0.0314	0.0314	0.9995	0.0314
1.9	0.0332	0.0332	0.9995	0.0332
2.0	0.0349	0.0349	0.9994	0.0349
2.1	0.0367	0.0366	0.9993	0.0367
2.2	0.0384	0.0384	0.9993	0.0384
2.3	0.0401	0.0401	0.9992	0.0402
2.4	0.0419	0.0419	0.9991	0.0419
2.5	0.0436	0.0436	0.9990	0.0437
2.6	0.0454	0.0454	0.9990	0.0454
2.7	0.0471	0.0471	0.9989	0.0472
2.8	0.0489	0.0488	0.9988	0.0489
2.9	0.0506	0.0506	0.9987	0.0507
3.0	0.0524	0.0523	0.9986	0.0524
3.1	0.0541	0.0541	0.9985	0.0542
3.2	0.0559	0.0558	0.9984	0.0559
3.3	0.0576	0.0576	0.9983	0.0577
3.4	0.0593	0.0593	0.9982	0.0594
3.5	0.0611	0.0610	0.9981	0.0612
3.6	0.0628	0.0628	0.9980	0.0629
3.7	0.0646	0.0645	0.9979	0.0647
3.8	0.0663	0.0663	0.9978	0.0664
3.9	0.0681	0.0680	0.9977	0.0682
4.0	0.0698	0.0698	0.9976	0.0699
4.1	0.0716	0.0715	0.9974	0.0717
4.2	0.0733	0.0732	0.9973	0.0734
4.3	0.0750	0.0750	0.9972	0.0752
4.4	0.0768	0.0767	0.9971	0.0769
4.5	0.0785	0.0785	0.9969	0.0787
4.6	0.0803	0.0802	0.9968	0.0805
4.7	0.0820	0.0819	0.9966	0.0822
4.8	0.0838	0.0837	0.9965	0.0840
4.9	0.0855	0.0854	0.9963	0.0857

TABLE I.   TRIGONOMETRIC TABLES—Continued

θ, Degrees	θ, Radians	Sin θ	Cos θ	Tan θ
5.0	0.0873	0.0872	0.9962	0.0875
5.1	0.0890	0.0889	0.9960	0.0892
5.2	0.0908	0.0906	0.9959	0.0910
5.3	0.0925	0.0924	0.9957	0.0928
5.4	0.0942	0.0941	0.9956	0.0945
5.5	0.0960	0.0958	0.9954	0.0963
5.6	0.0977	0.0976	0.9952	0.0981
5.7	0.0995	0.0993	0.9951	0.0998
5.8	0.1012	0.1011	0.9949	0.1016
5.9	0.1030	0.1028	0.9947	0.1033
6.0	0.1047	0.1045	0.9945	0.1051
6.1	0.1065	0.1063	0.9943	0.1069
6.2	0.1082	0.1080	0.9942	0.1086
6.3	0.1100	0.1097	0.9940	0.1104
6.4	0.1117	0.1115	0.9938	0.1122
6.5	0.1134	0.1132	0.9936	0.1139
6.6	0.1152	0.1149	0.9934	0.1157
6.7	0.1169	0.1167	0.9932	0.1175
6.8	0.1187	0.1184	0.9930	0.1192
6.9	0.1204	0.1201	0.9928	0.1210
7.0	0.1222	0.1219	0.9925	0.1228
7.1	0.1239	0.1236	0.9923	0.1246
7.2	0.1257	0.1253	0.9921	0.1263
7.3	0.1274	0.1271	0.9919	0.1281
7.4	0.1292	0.1288	0.9917	0.1299
7.5	0.1309	0.1305	0.9914	0.1317
7.6	0.1326	0.1323	0.9912	0.1334
7.7	0.1344	0.1340	0.9910	0.1352
7.8	0.1361	0.1357	0.9907	0.1370
7.9	0.1379	0.1374	0.9905	0.1388
8.0	0.1396	0.1392	0.9903	0.1405
8.1	0.1414	0.1409	0.9900	0.1423
8.2	0.1431	0.1426	0.9898	0.1441
8.3	0.1449	0.1444	0.9895	0.1459
8.4	0.1466	0.1461	0.9893	0.1477
8.5	0.1484	0.1478	0.9890	0.1495
8.6	0.1501	0.1495	0.9888	0.1512
8.7	0.1518	0.1513	0.9885	0.1530
8.8	0.1536	0.1530	0.9882	0.1548
8.9	0.1553	0.1547	0.9880	0.1566
9.0	0.1571	0.1564	0.9877	0.1584
9.1	0.1588	0.1582	0.9874	0.1602
9.2	0.1606	0.1599	0.9871	0.1620
9.3	0.1623	0.1616	0.9869	0.1638
9.4	0.1641	0.1633	0.9866	0.1655
9.5	0.1658	0.1650	0.9863	0.1673
9.6	0.1676	0.1668	0.9860	0.1691
9.7	0.1693	0.1685	0.9857	0.1709
9.8	0.1710	0.1702	0.9854	0.1727
9.9	0.1728	0.1719	0.9851	0.1745

TABLE I.   TRIGONOMETRIC TABLES—Continued

θ, Degrees	θ, Radians	Sin θ	Cos θ	Tan θ
10.0	0.1745	0.1736	0.9848	0.1763
10.1	0.1763	0.1754	0.9845	0.1781
10.2	0.1780	0.1771	0.9842	0.1799
10.3	0.1798	0.1788	0.9839	0.1817
10.4	0.1815	0.1805	0.9836	0.1835
10.5	0.1833	0.1822	0.9833	0.1853
10.6	0.1850	0.1840	0.9829	0.1871
10.7	0.1868	0.1857	0.9826	0.1890
10.8	0.1885	0.1874	0.9823	0.1908
10.9	0.1902	0.1891	0.9820	0.1926
11.0	0.1920	0.1908	0.9816	0.1944
11.1	0.1937	0.1925	0.9813	0.1962
11.2	0.1955	0.1942	0.9810	0.1980
11.3	0.1972	0.1959	0.9806	0.1998
11.4	0.1990	0.1977	0.9803	0.2016
11.5	0.2007	0.1994	0.9799	0.2035
11.6	0.2025	0.2011	0.9796	0.2053
11.7	0.2042	0.2028	0.9792	0.2071
11.8	0.2059	0.2045	0.9789	0.2089
11.9	0.2077	0.2062	0.9785	0.2107
12.0	0.2094	0.2079	0.9781	0.2126
12.1	0.2112	0.2096	0.9778	0.2144
12.2	0.2129	0.2113	0.9774	0.2162
12.3	0.2147	0.2130	0.9770	0.2180
12.4	0.2164	0.2147	0.9767	0.2199
12.5	0.2182	0.2164	0.9763	0.2217
12.6	0.2199	0.2181	0.9759	0.2235
12.7	0.2217	0.2198	0.9755	0.2254
12.8	0.2234	0.2215	0.9751	0.2272
12.9	0.2251	0.2232	0.9748	0.2290
13.0	0.2269	0.2250	0.9744	0.2309
13.1	0.2286	0.2267	0.9740	0.2327
13.2	0.2304	0.2284	0.9736	0.2345
13.3	0.2321	0.2300	0.9732	0.2364
13.4	0.2339	0.2317	0.9728	0.2382
13.5	0.2356	0.2334	0.9724	0.2401
13.6	0.2374	0.2351	0.9720	0.2419
13.7	0.2391	0.2368	0.9715	0.2438
13.8	0.2409	0.2385	0.9711	0.2456
13.9	0.2426	0.2402	0.9707	0.2475
14.0	0.2443	0.2419	0.9703	0.2493
14.1	0.2461	0.2436	0.9699	0.2512
14.2	0.2478	0.2453	0.9694	0.2530
14.3	0.2496	0.2470	0.9690	0.2549
14.4	0.2513	0.2487	0.9686	0.2568
14.5	0.2531	0.2504	0.9681	0.2586
14.6	0.2548	0.2521	0.9677	0.2605
14.7	0.2566	0.2538	0.9673	0.2623
14.8	0.2583	0.2554	0.9668	0.2642
14.9	0.2601	0.2571	0.9664	0.2661

TABLE I. TRIGONOMETRIC TABLES—Continued

θ, Degrees	θ, Radians	Sin θ	Cos θ	Tan θ
15.0	0.2618	0.2588	0.9659	0.2679
15.1	0.2635	0.2605	0.9655	0.2698
15.2	0.2653	0.2622	0.9650	0.2717
15.3	0.2670	0.2639	0.9646	0.2736
15.4	0.2688	0.2656	0.9641	0.2754
15.5	0.2705	0.2672	0.9636	0.2773
15.6	0.2723	0.2689	0.9632	0.2792
15.7	0.2740	0.2706	0.9627	0.2811
15.8	0.2758	0.2723	0.9622	0.2830
15.9	0.2775	0.2740	0.9617	0.2849
16.0	0.2793	0.2756	0.9613	0.2867
16.1	0.2810	0.2773	0.9608	0.2886
16.2	0.2827	0.2790	0.9603	0.2905
16.3	0.2845	0.2807	0.9598	0.2924
16.4	0.2862	0.2823	0.9593	0.2943
16.5	0.2880	0.2840	0.9588	0.2962
16.6	0.2897	0.2857	0.9583	0.2981
16.7	0.2915	0.2874	0.9578	0.3000
16.8	0.2932	0.2890	0.9573	0.3019
16.9	0.2950	0.2907	0.9568	0.3038
17.0	0.2967	0.2924	0.9563	0.3057
17.1	0.2985	0.2940	0.9558	0.3076
17.2	0.3002	0.2957	0.9553	0.3096
17.3	0.3019	0.2974	0.9548	0.3115
17.4	0.3037	0.2990	0.9542	0.3134
17.5	0.3054	0.3007	0.9537	0.3153
17.6	0.3072	0.3024	0.9532	0.3172
17.7	0.3089	0.3040	0.9527	0.3191
17.8	0.3107	0.3057	0.9521	0.3211
17.9	0.3124	0.3074	0.9516	0.3230
18.0	0.3142	0.3090	0.9511	0.3249
18.1	0.3159	0.3107	0.9505	0.3269
18.2	0.3176	0.3123	0.9500	0.3288
18.3	0.3194	0.3140	0.9494	0.3307
18.4	0.3211	0.3156	0.9489	0.3327
18.5	0.3229	0.3173	0.9483	0.3346
18.6	0.3246	0.3190	0.9478	0.3365
18.7	0.3264	0.3206	0.9472	0.3385
18.8	0.3281	0.3223	0.9466	0.3404
18.9	0.3299	0.3239	0.9461	0.3424
19.0	0.3316	0.3256	0.9455	0.3443
19.1	0.3334	0.3272	0.9449	0.3463
19.2	0.3351	0.3289	0.9444	0.3482
19.3	0.3368	0.3305	0.9438	0.3502
19.4	0.3386	0.3322	0.9432	0.3522
19.5	0.3403	0.3338	0.9426	0.3541
19.6	0.3421	0.3355	0.9421	0.3561
19.7	0.3438	0.3371	0.9415	0.3581
19.8	0.3456	0.3387	0.9409	0.3600
19.9	0.3473	0.3404	0.9403	0.3620

TABLE I. TRIGONOMETRIC TABLES—Continued

θ, Degrees	θ, Radians	Sin θ	Cos θ	Tan θ
20.0	0.3491	0.3420	0.9397	0.3640
20.1	0.3508	0.3437	0.9391	0.3659
20.2	0.3526	0.3453	0.9385	0.3679
20.3	0.3543	0.3469	0.9379	0.3699
20.4	0.3560	0.3486	0.9373	0.3719
20.5	0.3578	0.3502	0.9367	0.3739
20.6	0.3595	0.3518	0.9361	0.3759
20.7	0.3613	0.3535	0.9354	0.3779
20.8	0.3630	0.3551	0.9348	0.3799
20.9	0.3648	0.3567	0.9342	0.3819
21.0	0.3665	0.3584	0.9336	0.3839
21.1	0.3683	0.3600	0.9330	0.3859
21.2	0.3700	0.3616	0.9323	0.3879
21.3	0.3718	0.3633	0.9317	0.3899
21.4	0.3735	0.3649	0.9311	0.3919
21.5	0.3752	0.3665	0.9304	0.3939
21.6	0.3770	0.3681	0.9298	0.3959
21.7	0.3787	0.3697	0.9291	0.3979
21.8	0.3805	0.3714	0.9285	0.4000
21.9	0.3822	0.3730	0.9278	0.4020
22.0	0.3840	0.3746	0.9272	0.4040
22.1	0.3857	0.3762	0.9265	0.4061
22.2	0.3875	0.3778	0.9259	0.4081
22.3	0.3892	0.3795	0.9252	0.4101
22.4	0.3910	0.3811	0.9245	0.4122
22.5	0.3927	0.3827	0.9239	0.4142
22.6	0.3944	0.3843	0.9232	0.4163
22.7	0.3962	0.3859	0.9225	0.4183
22.8	0.3979	0.3875	0.9219	0.4204
22.9	0.3997	0.3891	0.9212	0.4224
23.0	0.4014	0.3907	0.9205	0.4245
23.1	0.4032	0.3923	0.9198	0.4265
23.2	0.4049	0.3939	0.9191	0.4286
23.3	0.4067	0.3955	0.9184	0.4307
23.4	0.4084	0.3971	0.9178	0.4327
23.5	0.4102	0.3987	0.9171	0.4348
23.6	0.4119	0.4003	0.9164	0.4369
23.7	0.4136	0.4019	0.9157	0.4390
23.8	0.4154	0.4035	0.9150	0.4411
23.9	0.4171	0.4051	0.9143	0.4431
24.0	0.4189	0.4067	0.9135	0.4452
24.1	0.4206	0.4083	0.9128	0.4473
24.2	0.4224	0.4099	0.9121	0.4494
24.3	0.4241	0.4115	0.9114	0.4515
24.4	0.4259	0.4131	0.9107	0.4536
24.5	0.4276	0.4147	0.9100	0.4557
24.6	0.4294	0.4163	0.9092	0.4578
24.7	0.4311	0.4179	0.9085	0.4599
24.8	0.4328	0.4195	0.9078	0.4621
24.9	0.4346	0.4210	0.9070	0.4642

TABLE I.   TRIGONOMETRIC TABLES—Continued

θ, Degrees	θ, Radians	Sin θ	Cos θ	Tan θ
25.0	0.4363	0.4226	0.9063	0.4663
25.1	0.4381	0.4242	0.9056	0.4684
25.2	0.4398	0.4258	0.9048	0.4706
25.3	0.4416	0.4274	0.9041	0.4727
25.4	0.4433	0.4289	0.9033	0.4748
25.5	0.4451	0.4305	0.9026	0.4770
25.6	0.4468	0.4321	0.9018	0.4791
25.7	0.4485	0.4337	0.9011	0.4813
25.8	0.4503	0.4352	0.9003	0.4834
25.9	0.4520	0.4368	0.8996	0.4856
26.0	0.4538	0.4384	0.8988	0.4877
26.1	0.4555	0.4399	0.8980	0.4899
26.2	0.4573	0.4415	0.8973	0.4921
26.3	0.4590	0.4431	0.8965	0.4942
26.4	0.4608	0.4446	0.8957	0.4964
26.5	0.4625	0.4462	0.8949	0.4986
26.6	0.4643	0.4478	0.8942	0.5008
26.7	0.4660	0.4493	0.8934	0.5029
26.8	0.4677	0.4509	0.8926	0.5051
26.9	0.4695	0.4524	0.8918	0.5073
27.0	0.4712	0.4540	0.8910	0.5095
27.1	0.4730	0.4555	0.8902	0.5117
27.2	0.4747	0.4571	0.8894	0.5139
27.3	0.4765	0.4586	0.8886	0.5161
27.4	0.4782	0.4602	0.8878	0.5184
27.5	0.4800	0.4617	0.8870	0.5206
27.6	0.4817	0.4633	0.8862	0.5228
27.7	0.4835	0.4648	0.8854	0.5250
27.8	0.4852	0.4664	0.8846	0.5272
27.9	0.4869	0.4679	0.8838	0.5295
28.0	0.4887	0.4695	0.8829	0.5317
28.1	0.4904	0.4710	0.8821	0.5339
28.2	0.4922	0.4726	0.8813	0.5362
28.3	0.4939	0.4741	0.8805	0.5384
28.4	0.4957	0.4756	0.8796	0.5407
28.5	0.4974	0.4772	0.8788	0.5430
28.6	0.4992	0.4787	0.8780	0.5452
28.7	0.5009	0.4802	0.8771	0.5475
28.8	0.5027	0.4818	0.8763	0.5498
28.9	0.5044	0.4833	0.8755	0.5520
29.0	0.5061	0.4848	0.8746	0.5543
29.1	0.5079	0.4863	0.8738	0.5566
29.2	0.5096	0.4879	0.8729	0.5589
29.3	0.5114	0.4894	0.8721	0.5612
29.4	0.5131	0.4909	0.8712	0.5635
29.5	0.5149	0.4924	0.8704	0.5658
29.6	0.5166	0.4939	0.8695	0.5681
29.7	0.5184	0.4955	0.8686	0.5704
29.8	0.5201	0.4970	0.8678	0.5727
29.9	0.5219	0.4985	0.8669	0.5750

TABLE I. TRIGONOMETRIC TABLES—Continued

θ, Degrees	θ, Radians	Sin θ	Cos θ	Tan θ
30.0	0.5236	0.5000	0.8660	0.5774
30.1	0.5253	0.5015	0.8652	0.5797
30.2	0.5271	0.5030	0.8643	0.5820
30.3	0.5288	0.5045	0.8634	0.5844
30.4	0.5306	0.5060	0.8625	0.5867
30.5	0.5323	0.5075	0.8616	0.5890
30.6	0.5341	0.5090	0.8607	0.5914
30.7	0.5358	0.5105	0.8599	0.5938
30.8	0.5376	0.5120	0.8590	0.5961
30.9	0.5393	0.5135	0.8581	0.5985
31.0	0.5411	0.5150	0.8572	0.6009
31.1	0.5428	0.5165	0.8563	0.6032
31.2	0.5445	0.5180	0.8554	0.6056
31.3	0.5463	0.5195	0.8545	0.6080
31.4	0.5480	0.5210	0.8536	0.6104
31.5	0.5498	0.5225	0.8526	0.6128
31.6	0.5515	0.5240	0.8517	0.6152
31.7	0.5533	0.5255	0.8508	0.6176
31.8	0.5550	0.5270	0.8499	0.6200
31.9	0.5568	0.5284	0.8490	0.6224
32.0	0.5585	0.5299	0.8480	0.6249
32.1	0.5603	0.5314	0.8471	0.6273
32.2	0.5620	0.5329	0.8462	0.6297
32.3	0.5637	0.5344	0.8453	0.6322
32.4	0.5655	0.5358	0.8443	0.6346
32.5	0.5672	0.5373	0.8434	0.6371
32.6	0.5690	0.5388	0.8425	0.6395
32.7	0.5707	0.5402	0.8415	0.6420
32.8	0.5725	0.5417	0.8406	0.6445
32.9	0.5742	0.5432	0.8396	0.6469
33.0	0.5760	0.5446	0.8387	0.6494
33.1	0.5777	0.5461	0.8377	0.6519
33.2	0.5794	0.5476	0.8368	0.6544
33.3	0.5812	0.5490	0.8358	0.6569
33.4	0.5829	0.5505	0.8348	0.6594
33.5	0.5847	0.5519	0.8339	0.6619
33.6	0.5864	0.5534	0.8329	0.6644
33.7	0.5882	0.5548	0.8320	0.6669
33.8	0.5899	0.5563	0.8310	0.6694
33.9	0.5917	0.5577	0.8300	0.6720
34.0	0.5934	0.5592	0.8290	0.6745
34.1	0.5952	0.5606	0.8281	0.6771
34.2	0.5969	0.5621	0.8271	0.6796
34.3	0.5986	0.5635	0.8261	0.6822
34.4	0.6004	0.5650	0.8251	0.6847
34.5	0.6021	0.5664	0.8241	0.6873
34.6	0.6039	0.5678	0.8231	0.6899
34.7	0.6056	0.5693	0.8221	0.6924
34.8	0.6074	0.5707	0.8211	0.6950
34.9	0.6091	0.5721	0.8202	0.6976

TABLE I. TRIGONOMETRIC TABLES—Continued

θ, Degrees	θ, Radians	Sin θ	Cos θ	Tan θ
35.0	0.6109	0.5736	0.8192	0.7002
35.1	0.6126	0.5750	0.8181	0.7028
35.2	0.6144	0.5764	0.8171	0.7054
35.3	0.6161	0.5779	0.8161	0.7080
35.4	0.6178	0.5793	0.8151	0.7107
35.5	0.6196	0.5807	0.8141	0.7133
35.6	0.6213	0.5821	0.8131	0.7159
35.7	0.6231	0.5835	0.8121	0.7186
35.8	0.6248	0.5850	0.8111	0.7212
35.9	0.6266	0.5864	0.8100	0.7239
36.0	0.6283	0.5878	0.8090	0.7265
36.1	0.6301	0.5892	0.8080	0.7292
36.2	0.6318	0.5906	0.8070	0.7319
36.3	0.6336	0.5920	0.8059	0.7346
36.4	0.6353	0.5934	0.8049	0.7373
36.5	0.6370	0.5948	0.8039	0.7400
36.6	0.6388	0.5962	0.8028	0.7427
36.7	0.6405	0.5976	0.8018	0.7454
36.8	0.6423	0.5990	0.8007	0.7481
36.9	0.6440	0.6004	0.7997	0.7508
37.0	0.6458	0.6018	0.7986	0.7536
37.1	0.6475	0.6032	0.7976	0.7563
37.2	0.6493	0.6046	0.7965	0.7590
37.3	0.6510	0.6060	0.7955	0.7618
37.4	0.6528	0.6074	0.7944	0.7646
37.5	0.6545	0.6088	0.7934	0.7673
37.6	0.6562	0.6101	0.7923	0.7701
37.7	0.6580	0.6115	0.7912	0.7729
37.8	0.6597	0.6129	0.7902	0.7757
37.9	0.6615	0.6143	0.7891	0.7785
38.0	0.6632	0.6157	0.7880	0.7813
38.1	0.6650	0.6170	0.7869	0.7841
38.2	0.6667	0.6184	0.7859	0.7869
38.3	0.6685	0.6198	0.7848	0.7898
38.4	0.6702	0.6211	0.7837	0.7926
38.5	0.6720	0.6225	0.7826	0.7954
38.6	0.6737	0.6239	0.7815	0.7983
38.7	0.6754	0.6252	0.7804	0.8012
38.8	0.6772	0.6266	0.7793	0.8040
38.9	0.6789	0.6280	0.7782	0.8069
39.0	0.6807	0.6293	0.7771	0.8098
39.1	0.6824	0.6307	0.7760	0.8127
39.2	0.6842	0.6320	0.7749	0.8156
39.3	0.6859	0.6334	0.7738	0.8185
39.4	0.6877	0.6347	0.7727	0.8214
39.5	0.6894	0.6361	0.7716	0.8243
39.6	0.6912	0.6374	0.7705	0.8273
39.7	0.6929	0.6388	0.7694	0.8302
39.8	0.6946	0.6401	0.7683	0.8332
39.9	0.6964	0.6414	0.7672	0.8361

TABLE I. TRIGONOMETRIC TABLES—Continued

θ, Degrees	θ, Radians	Sin θ	Cos θ	Tan θ
40.0	0.6981	0.6428	0.7660	0.8391
40.1	0.6999	0.6441	0.7649	0.8421
40.2	0.7016	0.6455	0.7638	0.8451
40.3	0.7034	0.6468	0.7627	0.8481
40.4	0.7051	0.6481	0.7615	0.8511
40.5	0.7069	0.6494	0.7604	0.8541
40.6	0.7086	0.6508	0.7593	0.8571
40.7	0.7103	0.6521	0.7581	0.8601
40.8	0.7121	0.6534	0.7570	0.8632
40.9	0.7138	0.6547	0.7559	0.8662
41.0	0.7156	0.6561	0.7547	0.8693
41.1	0.7173	0.6574	0.7536	0.8724
41.2	0.7191	0.6587	0.7524	0.8754
41.3	0.7208	0.6600	0.7513	0.8785
41.4	0.7226	0.6613	0.7501	0.8816
41.5	0.7243	0.6626	0.7490	0.8847
41.6	0.7261	0.6639	0.7478	0.8878
41.7	0.7278	0.6652	0.7466	0.8910
41.8	0.7295	0.6665	0.7455	0.8941
41.9	0.7313	0.6678	0.7443	0.8972
42.0	0.7330	0.6691	0.7431	0.9004
42.1	0.7348	0.6704	0.7420	0.9036
42.2	0.7365	0.6717	0.7408	0.9067
42.3	0.7383	0.6730	0.7396	0.9099
42.4	0.7400	0.6743	0.7385	0.9131
42.5	0.7418	0.6756	0.7373	0.9163
42.6	0.7435	0.6769	0.7361	0.9195
42.7	0.7453	0.6782	0.7349	0.9228
42.8	0.7470	0.6794	0.7337	0.9260
42.9	0.7487	0.6807	0.7325	0.9293
43.0	0.7505	0.6820	0.7314	0.9325
43.1	0.7522	0.6833	0.7302	0.9358
43.2	0.7540	0.6845	0.7290	0.9391
43.3	0.7557	0.6858	0.7278	0.9424
43.4	0.7575	0.6871	0.7266	0.9457
43.5	0.7592	0.6884	0.7254	0.9490
43.6	0.7610	0.6896	0.7242	0.9523
43.7	0.7627	0.6909	0.7230	0.9556
43.8	0.7645	0.6921	0.7218	0.9590
43.9	0.7662	0.6934	0.7206	0.9623
44.0	0.7679	0.6947	0.7193	0.9657
44.1	0.7697	0.6959	0.7181	0.9691
44.2	0.7714	0.6972	0.7169	0.9725
44.3	0.7732	0.6984	0.7157	0.9759
44.4	0.7749	0.6997	0.7145	0.9793
44.5	0.7767	0.7009	0.7133	0.9827
44.6	0.7784	0.7022	0.7120	0.9861
44.7	0.7802	0.7034	0.7108	0.9896
44.8	0.7819	0.7046	0.7096	0.9930
44.9	0.7837	0.7059	0.7083	0.9965

TABLE I. TRIGONOMETRIC TABLES – Continued

θ, Degrees	θ, Radians	Sin θ	Cos θ	Tan θ
45.0	0.7854	0.7071	0.7071	1.0000
45.1	0.7871	0.7083	0.7059	1.0035
45.2	0.7889	0.7096	0.7046	1.0070
45.3	0.7906	0.7108	0.7034	1.0105
45.4	0.7924	0.7120	0.7022	1.0141
45.5	0.7941	0.7133	0.7009	1.0176
45.6	0.7959	0.7145	0.6997	1.0212
45.7	0.7976	0.7157	0.6984	1.0247
45.8	0.7994	0.7169	0.6972	1.0283
45.9	0.8011	0.7181	0.6959	1.0319
46.0	0.8029	0.7193	0.6947	1.0355
46.1	0.8046	0.7206	0.6934	1.0392
46.2	0.8063	0.7218	0.6921	1.0428
46.3	0.8081	0.7230	0.6909	1.0464
46.4	0.8098	0.7242	0.6896	1.0501
46.5	0.8116	0.7254	0.6884	1.0538
46.6	0.8133	0.7266	0.6871	1.0575
46.7	0.8151	0.7278	0.6858	1.0612
46.8	0.8168	0.7290	0.6845	1.0649
46.9	0.8186	0.7302	0.6833	1.0686
47.0	0.8203	0.7314	0.6820	1.0724
47.1	0.8220	0.7325	0.6807	1.0761
47.2	0.8238	0.7337	0.6794	1.0799
47.3	0.8255	0.7349	0.6782	1.0837
47.4	0.8273	0.7361	0.6769	1.0875
47.5	0.8290	0.7373	0.6756	1.0913
47.6	0.8308	0.7385	0.6743	1.0951
47.7	0.8325	0.7396	0.6730	1.0990
47.8	0.8343	0.7408	0.6717	1.1028
47.9	0.8360	0.7420	0.6704	1.1067
48.0	0.8378	0.7431	0.6691	1.1106
48.1	0.8395	0.7443	0.6678	1.1145
48.2	0.8412	0.7455	0.6665	1.1184
48.3	0.8430	0.7466	0.6652	1.1224
48.4	0.8447	0.7478	0.6639	1.1263
48.5	0.8465	0.7490	0.6626	1.1303
48.6	0.8482	0.7501	0.6613	1.1343
48.7	0.8500	0.7513	0.6600	1.1383
48.8	0.8517	0.7524	0.6587	1.1423
48.9	0.8535	0.7536	0.6574	1.1463
49.0	0.8552	0.7547	0.6561	1.1504
49.1	0.8570	0.7559	0.6547	1.1544
49.2	0.8587	0.7570	0.6534	1.1585
49.3	0.8604	0.7581	0.6521	1.1626
49.4	0.8622	0.7593	0.6508	1.1667
49.5	0.8639	0.7604	0.6494	1.1708
49.6	0.8657	0.7615	0.6481	1.1750
49.7	0.8674	0.7627	0.6468	1.1792
49.8	0.8692	0.7638	0.6455	1.1833
49.9	0.8709	0.7649	0.6441	1.1875

TABLE I.   TRIGONOMETRIC TABLES—Continued

θ, Degrees	θ, Radians	Sin θ	Cos θ	Tan θ
50.0	0.8727	0.7660	0.6428	1.1918
50.1	0.8744	0.7672	0.6414	1.1960
50.2	0.8762	0.7683	0.6401	1.2002
50.3	0.8779	0.7694	0.6388	1.2045
50.4	0.8796	0.7705	0.6374	1.2088
50.5	0.8814	0.7716	0.6361	1.2131
50.6	0.8831	0.7727	0.6347	1.2174
50.7	0.8849	0.7738	0.6334	1.2218
50.8	0.8866	0.7749	0.6320	1.2261
50.9	0.8884	0.7760	0.6307	1.2305
51.0	0.8901	0.7771	0.6293	1.2349
51.1	0.8919	0.7782	0.6280	1.2393
51.2	0.8936	0.7793	0.6266	1.2437
51.3	0.8954	0.7804	0.6252	1.2482
51.4	0.8971	0.7815	0.6239	1.2527
51.5	0.8988	0.7826	0.6225	1.2572
51.6	0.9006	0.7837	0.6211	1.2617
51.7	0.9023	0.7848	0.6198	1.2662
51.8	0.9041	0.7859	0.6184	1.2708
51.9	0.9058	0.7869	0.6170	1.2753
52.0	0.9076	0.7880	0.6157	1.2799
52.1	0.9093	0.7891	0.6143	1.2846
52.2	0.9111	0.7902	0.6129	1.2892
52.3	0.9128	0.7912	0.6115	1.2938
52.4	0.9146	0.7923	0.6101	1.2985
52.5	0.9163	0.7934	0.6088	1.3032
52.6	0.9180	0.7944	0.6074	1.3079
52.7	0.9198	0.7955	0.6060	1.3127
52.8	0.9215	0.7965	0.6046	1.3174
52.9	0.9233	0.7976	0.6032	1.3222
53.0	0.9250	0.7986	0.6018	1.3270
53.1	0.9268	0.7997	0.6004	1.3319
53.2	0.9285	0.8007	0.5990	1.3367
53.3	0.9303	0.8018	0.5976	1.3416
53.4	0.9320	0.8028	0.5962	1.3465
53.5	0.9338	0.8039	0.5948	1.3514
53.6	0.9355	0.8049	0.5934	1.3564
53.7	0.9372	0.8059	0.5920	1.3613
53.8	0.9390	0.8070	0.5906	1.3663
53.9	0.9407	0.8080	0.5892	1.3713
54.0	0.9425	0.8090	0.5878	1.3764
54.1	0.9442	0.8100	0.5864	1.3814
54.2	0.9460	0.8111	0.5850	1.3865
54.3	0.9477	0.8121	0.5835	1.3916
54.4	0.9495	0.8131	0.5821	1.3968
54.5	0.9512	0.8141	0.5807	1.4019
54.6	0.9529	0.8151	0.5793	1.4071
54.7	0.9547	0.8161	0.5779	1.4123
54.8	0.9564	0.8171	0.5764	1.4176
54.9	0.9582	0.8181	0.5750	1.4229

TABLE I.   TRIGONOMETRIC TABLES—Continued

θ, Degrees	θ, Radians	Sin θ	Cos θ	Tan θ
55.0	0.9599	0.8192	0.5736	1.4281
55.1	0.9617	0.8202	0.5721	1.4335
55.2	0.9634	0.8211	0.5707	1.4388
55.3	0.9652	0.8221	0.5693	1.4442
55.4	0.9669	0.8231	0.5678	1.4496
55.5	0.9687	0.8241	0.5664	1.4550
55.6	0.9704	0.8251	0.5650	1.4605
55.7	0.9721	0.8261	0.5635	1.4659
55.8	0.9739	0.8271	0.5621	1.4715
55.9	0.9756	0.8281	0.5606	1.4770
56.0	0.9774	0.8290	0.5592	1.4826
56.1	0.9791	0.8300	0.5577	1.4882
56.2	0.9809	0.8310	0.5563	1.4938
56.3	0.9826	0.8320	0.5548	1.4994
56.4	0.9844	0.8329	0.5534	1.5051
56.5	0.9861	0.8339	0.5519	1.5108
56.6	0.9879	0.8348	0.5505	1.5166
56.7	0.9896	0.8358	0.5490	1.5224
56.8	0.9913	0.8368	0.5476	1.5282
56.9	0.9931	0.8377	0.5461	1.5340
57.0	0.9948	0.8387	0.5446	1.5399
57.1	0.9966	0.8396	0.5432	1.5458
57.2	0.9983	0.8406	0.5417	1.5517
57.3	1.0001	0.8415	0.5402	1.5577
57.4	1.0018	0.8425	0.5388	1.5637
57.5	1.0036	0.8434	0.5373	1.5697
57.6	1.0053	0.8443	0.5358	1.5757
57.7	1.0071	0.8453	0.5344	1.5818
57.8	1.0088	0.8462	0.5329	1.5880
57.9	1.0105	0.8471	0.5314	1.5941
58.0	1.0123	0.8480	0.5299	1.6003
58.1	1.0140	0.8490	0.5284	1.6066
58.2	1.0158	0.8499	0.5270	1.6128
58.3	1.0175	0.8508	0.5255	1.6191
58.4	1.0193	0.8517	0.5240	1.6255
58.5	1.0210	0.8526	0.5225	1.6318
58.6	1.0228	0.8536	0.5210	1.6383
58.7	1.0245	0.8545	0.5195	1.6447
58.8	1.0263	0.8554	0.5180	1.6512
58.9	1.0280	0.8563	0.5165	1.6577
59.0	1.0297	0.8572	0.5150	1.6643
59.1	1.0315	0.8581	0.5135	1.6709
59.2	1.0332	0.8590	0.5120	1.6775
59.3	1.0350	0.8599	0.5105	1.6842
59.4	1.0367	0.8607	0.5090	1.6909
59.5	1.0385	0.8616	0.5075	1.6977
59.6	1.0402	0.8625	0.5060	1.7045
59.7	1.0420	0.8634	0.5045	1.7113
59.8	1.0437	0.8643	0.5030	1.7182
59.9	1.0455	0.8652	0.5015	1.7251

TABLE I. TRIGONOMETRIC TABLES—Continued

θ. Degrees	θ, Radians	Sin θ	Cos θ	Tan θ
60.0	1.0472	0.8660	0.5000	1.7320
60.1	1.0489	0.8669	0.4985	1.7391
60.2	1.0507	0.8678	0.4970	1.7461
60.3	1.0524	0.8686	0.4955	1.7532
60.4	1.0542	0.8695	0.4939	1.7603
60.5	1.0559	0.8704	0.4924	1.7675
60.6	1.0577	0.8712	0.4909	1.7747
60.7	1.0594	0.8721	0.4894	1.7820
60.8	1.0612	0.8729	0.4879	1.7893
60.9	1.0629	0.8738	0.4863	1.7966
61.0	1.0646	0.8746	0.4848	1.8040
61.1	1.0664	0.8755	0.4833	1.8115
61.2	1.0681	0.8763	0.4818	1.8190
61.3	1.0699	0.8771	0.4802	1.8265
61.4	1.0716	0.8780	0.4787	1.8341
61.5	1.0734	0.8788	0.4772	1.8418
61.6	1.0751	0.8796	0.4756	1.8495
61.7	1.0769	0.8805	0.4741	1.8572
61.8	1.0786	0.8813	0.4726	1.8650
61.9	1.0804	0.8821	0.4710	1.8728
62.0	1.0821	0.8829	0.4695	1.8807
62.1	1.0838	0.8838	0.4679	1.8887
62.2	1.0856	0.8846	0.4664	1.8967
62.3	1.0873	0.8854	0.4648	1.9047
62.4	1.0891	0.8862	0.4633	1.9128
62.5	1.0908	0.8870	0.4617	1.9210
62.6	1.0926	0.8878	0.4602	1.9292
62.7	1.0943	0.8886	0.4586	1.9375
62.8	1.0961	0.8894	0.4571	1.9458
62.9	1.0978	0.8902	0.4555	1.9542
63.0	1.0996	0.8910	0.4540	1.9626
63.1	1.1013	0.8918	0.4524	1.9711
63.2	1.1030	0.8926	0.4509	1.9797
63.3	1.1048	0.8934	0.4493	1.9883
63.4	1.1065	0.8942	0.4478	1.9969
63.5	1.1083	0.8949	0.4462	2.0057
63.6	1.1100	0.8957	0.4446	2.0145
63.7	1.1118	0.8965	0.4431	2.0233
63.8	1.1135	0.8973	0.4415	2.0323
63.9	1.1153	0.8980	0.4399	2.0413
64.0	1.1170	0.8988	0.4384	2.0503
64.1	1.1188	0.8996	0.4368	2.0594
64.2	1.1205	0.9003	0.4352	2.0686
64.3	1.1222	0.9011	0.4337	2.0778
64.4	1.1240	0.9018	0.4321	2.0872
64.5	1.1257	0.9026	0.4305	2.0965
64.6	1.1275	0.9033	0.4289	2.1060
64.7	1.1292	0.9041	0.4274	2.1155
64.8	1.1310	0.9048	0.4258	2.1251
64.9	1.1327	0.9056	0.4242	2.1348

TABLE I. TRIGONOMETRIC TABLES—Continued

θ, Degrees	θ, Radians	Sin θ	Cos θ	Tan θ
65.0	1.1345	0.9063	0.4226	2.1445
65.1	1.1362	0.9070	0.4210	2.1543
65.2	1.1380	0.9078	0.4195	2.1642
65.3	1.1397	0.9085	0.4179	2.1741
65.4	1.1414	0.9092	0.4163	2.1842
65.5	1.1432	0.9100	0.4147	2.1943
65.6	1.1449	0.9107	0.4131	2.2045
65.7	1.1467	0.9114	0.4115	2.2148
65.8	1.1484	0.9121	0.4099	2.2251
65.9	1.1502	0.9128	0.4083	2.2355
66.0	1.1519	0.9135	0.4067	2.2460
66.1	1.1537	0.9143	0.4051	2.2566
66.2	1.1554	0.9150	0.4035	2.2673
66.3	1.1572	0.9157	0.4019	2.2781
66.4	1.1589	0.9164	0.4003	2.2889
66.5	1.1606	0.9171	0.3988	2.2998
66.6	1.1624	0.9178	0.3971	2.3109
66.7	1.1641	0.9184	0.3955	2.3220
66.8	1.1659	0.9191	0.3939	2.3332
66.9	1.1676	0.9198	0.3923	2.3445
67.0	1.1694	0.9205	0.3907	2.3558
67.1	1.1711	0.9212	0.3891	2.3673
67.2	1.1729	0.9219	0.3875	2.3789
67.3	1.1746	0.9225	0.3859	2.3906
67.4	1.1764	0.9232	0.3843	2.4023
67.5	1.1781	0.9239	0.3827	2.4142
67.6	1.1798	0.9245	0.3811	2.4262
67.7	1.1816	0.9252	0.3795	2.4382
67.8	1.1833	0.9259	0.3778	2.4504
67.9	1.1851	0.9265	0.3762	2.4627
68.0	1.1868	0.9272	0.3746	2.4751
68.1	1.1886	0.9278	0.3730	2.4876
68.2	1.1903	0.9285	0.3714	2.5002
68.3	1.1921	0.9291	0.3697	2.5129
68.4	1.1938	0.9298	0.3681	2.5257
68.5	1.1955	0.9304	0.3665	2.5386
68.6	1.1973	0.9311	0.3649	2.5517
68.7	1.1990	0.9317	0.3633	2.5649
68.8	1.2008	0.9323	0.3616	2.5781
68.9	1.2025	0.9330	0.3600	2.5916
69.0	1.2043	0.9336	0.3584	2.6051
69.1	1.2060	0.9342	0.3567	2.6187
69.2	1.2078	0.9348	0.3551	2.6325
69.3	1.2095	0.9354	0.3535	2.6464
69.4	1.2113	0.9361	0.3518	2.6605
69.5	1.2130	0.9367	0.3502	2.6746
69.6	1.2147	0.9373	0.3486	2.6889
69.7	1.2165	0.9379	0.3469	2.7033
69.8	1.2182	0.9385	0.3453	2.7179
69.9	1.2200	0.9391	0.3437	2.7326

TABLE I. TRIGONOMETRIC TABLES—Continued

θ, Degrees	θ, Radians	Sin θ	Cos θ	Tan θ
70.0	1.2217	0.9397	0.3420	2.7475
70.1	1.2235	0.9403	0.3404	2.7625
70.2	1.2252	0.9409	0.3387	2.7776
70.3	1.2270	0.9415	0.3371	2.7929
70.4	1.2287	0.9421	0.3355	2.8083
70.5	1.2305	0.9426	0.3338	2.8239
70.6	1.2322	0.9432	0.3322	2.8396
70.7	1.2339	0.9438	0.3305	2.8555
70.8	1.2357	0.9444	0.3289	2.8716
70.9	1.2374	0.9449	0.3272	2.8878
71.0	1.2392	0.9455	0.3256	2.9042
71.1	1.2409	0.9461	0.3239	2.9208
71.2	1.2427	0.9466	0.3223	2.9375
71.3	1.2444	0.9472	0.3206	2.9544
71.4	1.2462	0.9478	0.3190	2.9714
71.5	1.2479	0.9483	0.3173	2.9887
71.6	1.2497	0.9489	0.3156	3.0061
71.7	1.2514	0.9494	0.3140	3.0237
71.8	1.2531	0.9500	0.3123	3.0415
71.9	1.2549	0.9505	0.3107	3.0595
72.0	1.2566	0.9511	0.3090	3.0777
72.1	1.2584	0.9516	0.3074	3.0961
72.2	1.2601	0.9521	0.3057	3.1146
72.3	1.2619	0.9527	0.3040	3.1334
72.4	1.2636	0.9532	0.3024	3.1524
72.5	1.2654	0.9537	0.3007	3.1716
72.6	1.2671	0.9542	0.2990	3.1910
72.7	1.2689	0.9548	0.2974	3.2106
72.8	1.2706	0.9553	0.2957	3.2305
72.9	1.2723	0.9558	0.2940	3.2505
73.0	1.2741	0.9563	0.2924	3.2708
73.1	1.2758	0.9568	0.2907	3.2914
73.2	1.2776	0.9573	0.2890	3.3121
73.3	1.2793	0.9578	0.2874	3.3332
73.4	1.2811	0.9583	0.2857	3.3544
73.5	1.2828	0.9588	0.2840	3.3759
73.6	1.2846	0.9593	0.2823	3.3977
73.7	1.2863	0.9598	0.2807	3.4197
73.8	1.2881	0.9603	0.2790	3.4420
73.9	1.2898	0.9608	0.2773	3.4646
74.0	1.2915	0.9613	0.2756	3.4874
74.1	1.2933	0.9617	0.2740	3.5105
74.2	1.2950	0.9622	0.2723	3.5339
74.3	1.2968	0.9627	0.2706	3.5576
74.4	1.2985	0.9632	0.2689	3.5816
74.5	1.3003	0.9636	0.2672	3.6059
74.6	1.3020	0.9641	0.2656	3.6305
74.7	1.3038	0.9646	0.2639	3.6554
74.8	1.3055	0.9650	0.2622	3.6806
74.9	1.3073	0.9655	0.2605	3.7062

TABLE I. TRIGONOMETRIC TABLES—Continued

θ, Degrees	θ, Radians	Sin θ	Cos θ	Tan θ
75.0	1.3090	0.9659	0.2588	3.7320
75.1	1.3107	0.9664	0.2571	3.7583
75.2	1.3125	0.9668	0.2554	3.7848
75.3	1.3142	0.9673	0.2538	3.8118
75.4	1.3160	0.9677	0.2521	3.8390
75.5	1.3177	0.9681	0.2504	3.8667
75.6	1.3195	0.9686	0.2487	3.8947
75.7	1.3212	0.9690	0.2470	3.9231
75.8	1.3230	0.9694	0.2453	3.9520
75.9	1.3247	0.9699	0.2436	3.9812
76.0	1.3264	0.9703	0.2419	4.0108
76.1	1.3282	0.9707	0.2402	4.0408
76.2	1.3299	0.9711	0.2385	4.0712
76.3	1.3317	0.9715	0.2368	4.1021
76.4	1.3334	0.9720	0.2351	4.1335
76.5	1.3352	0.9724	0.2334	4.1653
76.6	1.3369	0.9728	0.2317	4.1975
76.7	1.3387	0.9732	0.2301	4.2303
76.8	1.3404	0.9736	0.2284	4.2635
76.9	1.3422	0.9740	0.2267	4.2972
77.0	1.3439	0.9744	0.2250	4.3315
77.1	1.3456	0.9748	0.2233	4.3662
77.2	1.3474	0.9751	0.2215	4.4015
77.3	1.3491	0.9755	0.2198	4.4373
77.4	1.3509	0.9759	0.2181	4.4737
77.5	1.3526	0.9763	0.2164	4.5107
77.6	1.3544	0.9767	0.2147	4.5482
77.7	1.3561	0.9770	0.2130	4.5864
77.8	1.3579	0.9774	0.2113	4.6252
77.9	1.3596	0.9778	0.2096	4.6646
78.0	1.3614	0.9781	0.2079	4.7046
78.1	1.3631	0.9785	0.2062	4.7453
78.2	1.3648	0.9789	0.2045	4.7867
78.3	1.3666	0.9792	0.2028	4.8288
78.4	1.3683	0.9796	0.2011	4.8716
78.5	1.3701	0.9799	0.1994	4.9151
78.6	1.3718	0.9803	0.1977	4.9594
78.7	1.3736	0.9806	0.1959	5.0045
78.8	1.3753	0.9810	0.1942	5.0503
78.9	1.3771	0.9813	0.1925	5.0970
79.0	1.3788	0.9816	0.1908	5.1445
79.1	1.3806	0.9820	0.1891	5.1929
79.2	1.3823	0.9823	0.1874	5.2422
79.3	1.3840	0.9826	0.1857	5.2923
79.4	1.3858	0.9829	0.1840	5.3434
79.5	1.3875	0.9833	0.1822	5.3955
79.6	1.3893	0.9836	0.1805	5.4485
79.7	1.3910	0.9839	0.1788	5.5026
79.8	1.3928	0.9842	0.1771	5.5577
79.9	1.3945	0.9845	0.1754	5.6139

TABLE I. TRIGONOMETRIC TABLES—Continued

θ. Degrees	θ, Radians	Sin θ	Cos θ	Tan θ
80.0	1.3963	0.9848	0.1736	5.6713
80.1	1.3980	0.9851	0.1719	5.7297
80.2	1.3998	0.9854	0.1702	5.7894
80.3	1.4015	0.9857	0.1685	5.8502
80.4	1.4032	0.9860	0.1668	5.9123
80.5	1.4050	0.9863	0.1650	5.9757
80.6	1.4067	0.9866	0.1633	6.0405
80.7	1.4085	0.9869	0.1616	6.1066
80.8	1.4102	0.9871	0.1599	6.1741
80.9	1.4120	0.9874	0.1582	6.2432
81.0	1.4137	0.9877	0.1564	6.3138
81.1	1.4155	0.9880	0.1547	6.3859
81.2	1.4172	0.9882	0.1530	6.4596
81.3	1.4190	0.9885	0.1513	6.5350
81.4	1.4207	0.9888	0.1495	6.6122
81.5	1.4224	0.9890	0.1478	6.6912
81.6	1.4242	0.9893	0.1461	6.7720
81.7	1.4259	0.9895	0.1444	6.8548
81.8	1.4277	0.9898	0.1426	6.9395
81.9	1.4294	0.9900	0.1409	7.0264
82.0	1.4312	0.9903	0.1392	7.1154
82.1	1.4329	0.9905	0.1374	7.2066
82.2	1.4347	0.9907	0.1357	7.3002
82.3	1.4364	0.9910	0.1340	7.3962
82.4	1.4382	0.9912	0.1323	7.4947
82.5	1.4399	0.9914	0.1305	7.5958
82.6	1.4416	0.9917	0.1288	7.6996
82.7	1.4434	0.9919	0.1271	7.8062
82.8	1.4451	0.9921	0.1253	7.9158
82.9	1.4469	0.9923	0.1236	8.0285
83.0	1.4486	0.9925	0.1219	8.1443
83.1	1.4504	0.9928	0.1201	8.2636
83.2	1.4521	0.9930	0.1184	8.3863
83.3	1.4539	0.9932	0.1167	8.5126
83.4	1.4556	0.9934	0.1149	8.6427
83.5	1.4573	0.9936	0.1132	8.7769
83.6	1.4591	0.9938	0.1115	8.9152
83.7	1.4608	0.9940	0.1097	9.0579
83.8	1.4626	0.9942	0.1080	9.2052
83.9	1.4643	0.9943	0.1063	9.3572
84.0	1.4661	0.9945	0.1045	9.5144
84.1	1.4678	0.9947	0.1028	9.6768
84.2	1.4696	0.9949	0.1011	9.8448
84.3	1.4713	0.9951	0.0993	10.019
84.4	1.4731	0.9952	0.0976	10.199
84.5	1.4748	0.9954	0.0958	10.385
84.6	1.4765	0.9956	0.0941	10.579
84.7	1.4783	0.9957	0.0924	10.780
84.8	1.4800	0.9959	0.0906	10.988
84.9	1.4818	0.9960	0.0889	11.205

TABLE I.   TRIGONOMETRIC TABLES—Continued

θ, Degrees	θ, Radians	Sin θ	Cos θ	Tan θ
85.0	1.4835	0.9962	0.0872	11.430
85.1	1.4853	0.9963	0.0854	11.664
85.2	1.4870	0.9965	0.0837	11.909
85.3	1.4888	0.9966	0.0819	12.163
85.4	1.4905	0.9968	0.0802	12.429
85.5	1.4923	0.9969	0.0785	12.706
85.6	1.4940	0.9971	0.0767	12.996
85.7	1.4957	0.9972	0.0750	13.300
85.8	1.4975	0.9973	0.0732	13.617
85.9	1.4992	0.9974	0.0715	13.951
86.0	1.5010	0.9976	0.0698	14.301
86.1	1.5027	0.9977	0.0680	14.669
86.2	1.5045	0.9978	0.0663	15.056
86.3	1.5062	0.9979	0.0645	15.464
86.4	1.5080	0.9980	0.0628	15.895
86.5	1.5097	0.9981	0.0610	16.350
86.6	1.5115	0.9982	0.0593	16.832
86.7	1.5132	0.9983	0.0576	17.343
86.8	1.5149	0.9984	0.0558	17.886
86.9	1.5167	0.9985	0.0541	18.464
87.0	1.5184	0.9986	0.0523	19.081
87.1	1.5202	0.9987	0.0506	19.740
87.2	1.5219	0.9988	0.0489	20.446
87.3	1.5237	0.9989	0.0471	21.205
87.4	1.5254	0.9990	0.0454	22.022
87.5	1.5272	0.9990	0.0436	22.904
87.6	1.5289	0.9991	0.0419	23.859
87.7	1.5307	0.9992	0.0401	24.898
87.8	1.5324	0.9993	0.0384	26.031
87.9	1.5341	0.9993	0.0366	27.271
88.0	1.5359	0.9994	0.0349	28.636
88.1	1.5376	0.9995	0.0332	30.145
88.2	1.5394	0.9995	0.0314	31.821
88.3	1.5411	0.9996	0.0297	33.694
88.4	1.5429	0.9996	0.0279	35.801
88.5	1.5446	0.9997	0.0262	38.188
88.6	1.5464	0.9997	0.0244	40.917
88.7	1.5481	0.9997	0.0227	44.066
88.8	1.5499	0.9998	0.0209	47.740
88.9	1.5516	0.9998	0.0192	52.882
89.0	1.5533	0.9998	0.0175	57.290
89.1	1.5551	0.9999	0.0157	63.657
89.2	1.5568	0.9999	0.0140	71.615
89.3	1.5586	0.9999	0.0122	81.847
89.4	1.5603	0.9999	0.0105	95.489
89.5	1.5621	1.0000	0.0087	114.59
89.6	1.5638	1.0000	0.0070	143.24
89.7	1.5656	1.0000	0.0052	190.98
89.8	1.5673	1.0000	0.0035	286.48
89.9	1.5690	1.0000	0.0017	572.96
90.0	1.5708	1.0000	0.0000	∞

See following page for Table II

TABLE II. SQUARES, CUBES, AND ROOTS*

n'	n^2	$\sqrt{n}$	$\sqrt{10n}$	n^3	$\sqrt[3]{n}$	$\sqrt[3]{10n}$	$\sqrt[3]{100n}$
1	1	1.000 000	3.162 278	1	1.000 000	2.154 435	4.641 589
2	4	1.414 214	4.472 136	8	1.259 921	2.714 418	5.848 035
3	9	1.732 051	5.477 226	27	1.442 250	3.107 233	6.694 330
4	16	2.000 000	6.324 555	64	1.587 401	3.419 952	7.368 063
5	25	2.236 068	7.071 068	125	1.709 976	3.684 031	7.937 005
6	36	2.449 490	7.745 967	216	1.817 121	3.914 868	8.434 327
7	49	2.645 751	8.366 600	343	1.912 931	4.121 285	8.879 040
8	64	2.828 427	8.944 272	512	2.000 000	4.308 869	9.283 178
9	81	3.000 000	9.486 833	729	2.080 084	4.481 405	9.654 894
10	100	3.162 278	10.00000	1 000	2.154 435	4.641 589	10.00000
11	121	3.316 625	10.48809	1 331	2.223 980	4.791 420	10.32280
12	144	3.464 102	10.95445	1 728	2.289 428	4.932 424	10.62659
13	169	3.605 551	11.40175	2 197	2.351 335	5.065 797	10.91393
14	196	3.741 657	11.83216	2 744	2.410 142	5.192 494	11.18689
15	225	3.872 983	12.24745	3 375	2.466 212	5.313 293	11.44714
16	256	4.000 000	12.64911	4 096	2.519 842	5.428 835	11.69607
17	289	4.123 106	13.03840	4 913	2.571 282	5.539 658	11.93483
18	324	4.242 641	13.41641	5 832	2.620 741	5.646 216	12.16440
19	361	4.358 899	13.78405	6 859	2.668 402	5.748 897	12.38562
20	400	4.472 136	14.14214	8 000	2.714 418	5.848 035	12.59921
21	441	4.582 576	14.49138	9 261	2.758 924	5.943 922	12.80579
22	484	4.690 416	14.83240	10 648	2.802 039	6.036 811	13.00591
23	529	4.795 832	15.16575	12 167	2.843 867	6.126 926	13.20006
24	576	4.898 979	15.49193	13 824	2.884 499	6.214 465	13.38866
25	625	5.000 000	15.81139	15 625	2.924 018	6.299 605	13.57209
26	676	5.099 020	16.12452	17 576	2.962 496	6.382 504	13.75069
27	729	5.196 152	16.43168	19 683	3.000 000	6.463 304	13.92477
28	784	5.291 503	16.73320	21 952	3.036 589	6.542 133	14.09460
29	841	5.385 165	17.02939	24 389	3.072 317	6.619 106	14.26043
30	900	5.477 226	17.32051	27 000	3.107 233	6.694 330	14.42250
31	961	5.567 764	17.60682	29 791	3.141 381	6.767 899	14.58100
32	1 024	5.656 854	17.88854	32 768	3.174 802	6.839 904	14.73613
33	1 089	5.744 563	18.16590	35 937	3.207 534	6.910 423	14.88806
34	1 156	5.830 952	18.43909	39 304	3.239 612	6.979 532	15.03695
35	1 225	5.916 080	18.70829	42 875	3.271 066	7.047 299	15.18294
36	1 296	6.000 000	18.97367	46 656	3.301 927	7.113 787	15.32619
37	1 369	6.082 763	19.23538	50 653	3.332 222	7.179 054	15.46680
38	1 444	6.164 414	19.49359	54 872	3.361 975	7.243 156	15.60491
39	1 521	6.244 998	19.74842	59 319	3.391 211	7.306 144	15.74061
40	1 600	6.324 555	20.00000	64 000	3.419 952	7.368 063	15.87401
41	1 681	6.403 124	20.24846	68 921	3.448 217	7.428 959	16.00521
42	1 764	6.480 741	20.49390	74 088	3.476 027	7.488 872	16.13429
43	1 849	6.557 439	20.73644	79 507	3.503 398	7.547 842	16.26133
44	1 936	6.633 250	20.97618	85 184	3.530 348	7.605 905	16.38643
45	2 025	6.708 204	21.21320	91 125	3.556 893	7.663 094	16.50964
46	2 116	6.782 330	21.44761	97 336	3.583 048	7.719 443	16.63103
47	2 209	6.855 655	21.67948	103 823	3.608 826	7.774 980	16.75069
48	2 304	6.928 203	21.90890	110 592	3.634 241	7.829 735	16.86865
49	2 401	7.000 000	22.13594	117 649	3.659 306	7.883 735	16.98499
50	2 500	7.071 068	22.36068	125 000	3.684 031	7.937 005	17.09976

*Roots of numbers other than those given in this table may be determined from the following relations:

Square Roots:

$$\sqrt{1000n} = 10\sqrt{10n}; \quad \sqrt{100n} = 10\sqrt{n}; \quad \sqrt{\frac{n}{10}} = \frac{\sqrt{10n}}{10}; \quad \sqrt{\frac{n}{100}} = \frac{\sqrt{n}}{10}; \quad \sqrt{\frac{n}{1000}} = \frac{\sqrt{10n}}{100}$$

Cube Roots:

$$\sqrt[3]{100,000n} = 10\sqrt[3]{100n}; \quad \sqrt[3]{10,000n} = 10\sqrt[3]{10n}; \quad \sqrt[3]{1000n} = 10\sqrt[3]{n}; \quad \sqrt[3]{\frac{n}{100}} = \frac{\sqrt[3]{10n}}{10}; \quad \sqrt[3]{\frac{n}{1000}} = \frac{\sqrt[3]{n}}{10}$$

TABLE II. SQUARES, CUBES, AND ROOTS—Continued

n'	n^2	$\sqrt{n}$	$\sqrt{10n}$	n^3	$\sqrt[3]{n}$	$\sqrt[3]{10n}$	$\sqrt[3]{100n}$
50	2 500	7.071 068	22.36068	125 000	3.684 031	7.937 005	17.09976
51	2 601	7.141 428	22.58318	132 651	3.708 430	7.989 570	17.21301
52	2 704	7.211 103	22.80351	140 608	3.732 511	8.041 452	17.32478
53	2 809	7.280 110	23.02173	148 877	3.756 286	8.092 672	17.43513
54	2 916	7.348 469	23.23790	157 464	3.779 763	8.143 253	17.54411
55	3 025	7.416 198	23.45208	166 375	3.802 952	8.193 213	17.65174
56	3 136	7.483 315	23.66432	175 616	3.825 862	8.242 571	17.75808
57	3 249	7.549 834	23.87467	185 193	3.848 501	8.291 344	17.86316
58	3 364	7.615 773	24.08319	195 112	3.870 877	8.339 551	17.96702
59	3 481	7.681 146	24.28992	205 379	3.892 996	8.387 207	18.06969
60	3 600	7.745 967	24.49490	216 000	3.914 868	8.434 327	18.17121
61	3 721	7.810 250	24.69818	226 981	3.936 497	8.480 926	18.27160
62	3 844	7.874 008	24.89980	238 328	3.957 892	8.527 019	18.37091
63	3 969	7.937 254	25.09980	250 047	3.979 057	8.572 619	18.46915
64	4 096	8.000 000	25.29822	262 144	4.000 000	8.617 739	18.56636
65	4 225	8.062 258	25.49510	274 625	4.020 726	8.662 391	18.66256
66	4 356	8.124 038	25.69047	287 496	4.041 240	8.706 588	18.75777
67	4 489	8.185 353	25.88436	300 763	4.061 548	8.750 340	18.85204
68	4 624	8.246 211	26.07681	314 432	4.081 655	8.793 659	18.94536
69	4 761	8.306 624	26.26785	328 509	4.101 566	8.836 556	19.03778
70	4 900	8.366 600	26.45751	343 000	4.121 285	8.879 040	19.12931
71	5 041	8.426 150	26.64583	357 911	4.140 818	8.921 121	19.21997
72	5 184	8.485 281	26.83282	373 248	4.160 168	8.962 809	19.30979
73	5 329	8.544 004	27.01851	389 017	4.179 339	9.004 113	19.39877
74	5 476	8.602 325	27.20294	405 224	4.198 336	9.045 042	19.48695
75	5 625	8.660 254	27.38613	421 875	4.217 163	9.085 603	19.57434
76	5 776	8.717 798	27.56810	438 976	4.235 824	9.125 805	19.66095
77	5 929	8.774 964	27.74887	456 533	4.254 321	9.165 656	19.74681
78	6 084	8.831 761	27.92848	474 552	4.272 659	9.205 164	19.83192
79	6 241	8.888 194	28.10694	493 039	4.290 840	9.244 335	19.91632
80	6 400	8.944 272	28.28427	512 000	4.308 869	9.283 178	20.00000
81	6 561	9.000 000	28.46050	531 441	4.326 749	9.321 698	20.08299
82	6 724	9.055 385	28.63564	551 368	4.344 481	9.359 902	20.16530
83	6 889	9.110 434	28.80972	571 787	4.362 071	9.397 796	20.24694
84	7 056	9.165 151	28.98275	592 704	4.379 519	9.435 388	20.32793
85	7 225	9.219 544	29.15476	614 125	4.396 830	9.472 682	20.40828
86	7 396	9.273 618	29.32576	636 056	4.414 005	9.509 685	20.48800
87	7 569	9.327 379	29.49576	658 503	4.431 048	9.546 403	20.56710
88	7 744	9.380 832	29.66479	681 472	4.447 960	9.582 840	20.64560
89	7 921	9.433 981	29.83287	704 969	4.464 745	9.619 002	20.72351
90	8 100	9.486 833	30.00000	729 000	4.481 405	9.654 894	20.80084
91	8 281	9.539 392	30.16621	753 571	4.497 941	9.690 521	20.87759
92	8 464	9.591 663	30.33150	778 688	4.514 357	9.725 888	20.95379
93	8 649	9.643 651	30.49590	804 357	4.530 655	9.761 000	21.02944
94	8 836	9.695 360	30.65942	830 584	4.546 836	9.795 861	21.10454
95	9 025	9.746 794	30.82207	857 375	4.562 903	9.830 476	21.17912
96	9 216	9.797 959	30.98387	884 736	4.578 857	9.864 848	21.25317
97	9 409	9.848 858	31.14482	912 673	4.594 701	9.898 983	21.32671
98	9 604	9.899 495	31.30495	941 192	4.610 436	9.932 884	21.39975
99	9 801	9.949 874	31.46427	970 299	4.626 065	9.966 555	21.47229
100	10 000	10.00000	31.62278	1 000 000	4.641 589	10.00000	21.54435

TABLE III. LOGARITHMS TO THE BASE 10

The common logarithms[1] of numbers other than those given in the table below may be determined from the following relations.[2]

$$\log 10n = \log n + 1; \ \log 100n = \log n + 2; \ \log 1000n = \log n + 3$$
$$\log \frac{n}{10} = \log n - 1; \ \log \frac{n}{100} = \log n - 2; \ \log \frac{n}{1000} = \log n - 3.$$

n	log n	n	log n	n	log n
1.0	.0000	4.0	.6021	7.0	.8451
1.1	.0414	4.1	.6128	7.1	.8513
1.2	.0792	4.2	.6232	7.2	.8573
1.3	.1139	4.3	.6335	7.3	.8633
1.4	.1461	4.4	.6435	7.4	.8692
1.5	.1761	4.5	.6532	7.5	.8751
1.6	.2041	4.6	.6628	7.6	.8808
1.7	.2304	4.7	.6721	7.7	.8865
1.8	.2553	4.8	.6812	7.8	.8921
1.9	.2788	4.9	.6902	7.9	.8976
2.0	.3010	5.0	.6990	8.0	.9031
2.1	.3222	5.1	.7076	8.1	.9085
2.2	.3424	5.2	.7160	8.2	.9138
2.3	.3617	5.3	.7243	8.3	.9191
2.4	.3802	5.4	.7324	8.4	.9243
2.5	.3979	5.5	.7404	8.5	.9294
2.6	.4150	5.6	.7482	8.6	.9345
2.7	.4314	5.7	.7559	8.7	.9395
2.8	.4472	5.8	.7634	8.8	.9445
2.9	.4624	5.9	.7709	8.9	.9494
3.0	.4771	6.0	.7782	9.0	.9542
3.1	.4914	6.1	.7853	9.1	.9590
3.2	.5051	6.2	.7924	9.2	.9638
3.3	.5185	6.3	.7993	9.3	.9685
3.4	.5315	6.4	.8062	9.4	.9731
3.5	.5441	6.5	.8129	9.5	.9777
3.6	.5563	6.6	.8195	9.6	.9823
3.7	.5682	6.7	.8261	9.7	.9868
3.8	.5798	6.8	.8325	9.8	.9912
3.9	.5911	6.9	.8388	9.9	.9956

[1] Logarithms to the base 10.

[2] These relations follow from the identities $\log AB = \log A + \log B$, and $\log \frac{A}{B} = \log A - \log B$.

TABLE IV. THE GREEK ALPHABET

Letter		Name	Letter		Name
A α		alpha	N ν		nu
B β		beta	Ξ ξ		xi
Γ γ		gamma	O o		omicron
Δ δ		delta	Π π		pi
E ϵ		epsilon	P ρ		rho
Z ζ		zeta	Σ σ		sigma
H η		eta	T τ		tau
Θ θ ϑ		theta	Y υ		upsilon
I ι		iota	Φ ϕ φ		phi
K κ		kappa	X χ		chi
Λ λ		lambda	Ψ ψ		psi
M μ		mu	Ω ω		omega

TABLE V.   PERIODIC TABLE OF THE ELEMENTS

GROUPS

| | IA | IIA | IIIB | IVB | VB | VIB | VIIB | | VIII | | IB | IIB | IIIA | IVA | VA | VIA | VIIA | O |

PERIODS

Period 1:
1 H 1.00797 ; 2 He 4.0026

Period 2:
3 Li 6.939 ; 4 Be 9.0122 ; 5 B 10.811 ; 6 C 12.01115 ; 7 N 14.0067 ; 8 O 15.9994 ; 9 F 18.9984 ; 10 Ne 20.183

Period 3:
11 Na 22.9898 ; 12 Mg 24.312 ; 13 Al 26.9815 ; 14 Si 28.086 ; 15 P 30.9738 ; 16 S 32.064 ; 17 Cl 35.453 ; 18 Ar 39.948

Period 4:
19 K 39.102 ; 20 Ca 40.08 ; 21 Sc 44.956 ; 22 Ti 47.90 ; 23 V 50.942 ; 24 Cr 51.996 ; 25 Mn 54.9380 ; 26 Fe 55.847 ; 27 Co 58.9332 ; 28 Ni 58.71 ; 29 Cu 63.54 ; 30 Zn 65.37 ; 31 Ga 69.72 ; 32 Ge 72.59 ; 33 As 74.9216 ; 34 Se 78.96 ; 35 Br 79.909 ; 36 Kr 83.80

Period 5:
37 Rb 85.47 ; 38 Sr 87.62 ; 39 Y 88.905 ; 40 Zr 91.22 ; 41 Nb 92.906 ; 42 Mo 95.94 ; 43 Tc (99) ; 44 Ru 101.07 ; 45 Rh 102.905 ; 46 Pd 106.4 ; 47 Ag 107.870 ; 48 Cd 112.40 ; 49 In 114.82 ; 50 Sn 118.69 ; 51 Sb 121.75 ; 52 Te 127.60 ; 53 I 126.9044 ; 54 Xe 131.30

Period 6:
55 Cs 132.905 ; 56 Ba 137.34 ; 57 La 138.91 ; 58 ▶ 71 Ce ▶ Lu ; 72 Hf 178.49 ; 73 Ta 180.948 ; 74 W 183.85 ; 75 Re 186.2 ; 76 Os 190.2 ; 77 Ir 192.2 ; 78 Pt 195.09 ; 79 Au 196.967 ; 80 Hg 200.59 ; 81 Tl 204.37 ; 82 Pb 207.19 ; 83 Bi 208.980 ; 84 Po (210) ; 85 At (210) ; 86 Rn (222)

Period 7:
87 Fr (223) ; 88 Ra (226) ; 89 Ac (227) ; 90 ▶ 103 Th ▶ Lr ; 104 Ku (260) ; 105 Ha (260) ; 106 ; 107

LANTHANIDE SERIES:
58 Ce 140.12 ; 59 Pr 140.907 ; 60 Nd 144.24 ; 61 Pm (147) ; 62 Sm 150.35 ; 63 Eu 151.96 ; 64 Gd 157.25 ; 65 Tb 158.924 ; 66 Dy 162.50 ; 67 Ho 164.930 ; 68 Er 167.26 ; 69 Tm 168.934 ; 70 Yb 173.04 ; 71 Lu 174.97

ACTINIDE SERIES:
90 Th 232.038 ; 91 Pa (231) ; 92 U 238.03 ; 93 Np (237) ; 94 Pu (242) ; 95 Am (243) ; 96 Cm (247) ; 97 Bk (249) ; 98 Cf (251) ; 99 Es (254) ; 100 Fm (256) ; 101 Md (256) ; 102 No (254) ; 103 Lw (257)

Key:

| 26 |
| Fe |
| 55.847 |

Atomic number (Z) (= number of protons in nucleus)

Element symbol

Atomic mass (in AMU) of the naturally occurring isotopic mixture. $1 \text{ AMU} = 1.6605 \times 10^{-27}$ kg $= (1/12) \times$ (mass of C^{12} atom).

For the elements that are naturally radioactive, the numbers in parentheses are mass numbers of the most stable isotope of those elements.

TABLE VI. ATOMIC MASSES OF THE NATURALLY OCCURRING ISOTOPIC MIXTURES OF THE ELEMENTS

Element	Symbol	Atomic No.	Atomic Mass (AMU)†	Element	Symbol	Atomic No.	Atomic Mass (AMU)†
Actinium	Ac	89	[227]*	Mercury	Hg	80	200.59
Aluminum	Al	13	26.9815	Molybdenum	Mo	42	95.94
Americium	Am	95	[243]*	Neodymium	Nd	60	144.24
Antimony	Sb	51	121.75	Neon	Ne	10	20.183
Argon	Ar	18	39.948	Neptunium	Np	93	[237]*
Arsenic	As	33	74.9216	Nickel	Ni	28	58.71
Astatine	At	85	[210]*	Niobium	Nb	41	92.906
Barium	Ba	56	137.34	Nitrogen	N	7	14.0067
Berkelium	Bk	97	[247]*	Nobelium	No	102	[253]*
Beryllium	Be	4	9.0122	Osmium	Os	76	190.2
Bismuth	Bi	83	208.980	Oxygen	O	8	15.9994
Boron	B	5	10.811	Palladium	Pd	46	106.4
Bromine	Br	35	79.909	Phosphorus	P	15	30.9738
Cadmium	Cd	48	112.40	Platinum	Pt	78	195.09
Calcium	Ca	20	40.08	Plutonium	Pu	94	[242]*
Californium	Cf	98	[249]*	Polonium	Po	84	[210]*
Carbon	C	6	12.01115	Potassium	K	19	39.102
Cerium	Ce	58	140.12	Praseodymium	Pr	59	140.907
Cesium	Cs	55	132.905	Promethium	Pm	61	[145]*
Chlorine	Cl	17	35.453	Protactinium	Pa	91	[231]*
Chromium	Cr	24	51.996	Radium	Ra	88	[226.05]*
Cobalt	Co	27	58.9332	Radon	Rn	86	[222]*
Copper	Cu	29	63.54	Rhenium	Re	75	186.2
Curium	Cm	96	[248]*	Rhodium	Rh	45	102.905
Dysprosium	Dy	66	162.50	Rubidium	Rb	37	85.47
Einsteinium	Es	99	[254]*	Ruthenium	Ru	44	101.07
Erbium	Er	68	167.26	Samarium	Sm	62	150.35
Europium	Eu	63	151.96	Scandium	Sc	21	44.956
Fermium	Fm	100	[253]*	Selenium	Se	34	78.96
Fluorine	F	9	18.9984	Silicon	Si	14	28.086
Francium	Fr	87	[223]*	Silver	Ag	47	107.870
Gadolinium	Gd	64	157.25	Sodium	Na	11	22.9898
Gallium	Ga	31	69.72	Strontium	Sr	38	87.62
Germanium	Ge	32	72.59	Sulfur	S	16	32.064
Gold	Au	79	196.967	Tantalum	Ta	73	180.948
Hafnium	Hf	72	178.49	Technetium	Tc	43	[99]*
Helium	He	2	4.0026	Tellurium	Te	52	127.60
Holmium	Ho	67	164.930	Terbium	Tb	65	158.924
Hydrogen	H	1	1.00797	Thallium	Tl	81	204.37
Indium	In	49	114.82	Thorium	Th	90	232.038
Iodine	I	53	126.9044	Thulium	Tm	69	168.934
Iridium	Ir	77	192.2	Tin	Sn	50	118.69
Iron	Fe	26	55.847	Titanium	Ti	22	47.90
Krypton	Kr	36	83.80	Tungsten	W	74	183.85
Lanthanum	La	57	138.91	Uranium	U	92	238.03
Lawrencium	Lw	103	[259]*	Vanadium	V	23	50.942
Lead	Pb	82	207.19	Xenon	Xe	54	131.30
Lithium	Li	3	6.939	Ytterbium	Yb	70	173.04
Lutetium	Lu	71	174.97	Yttrium	Y	39	88.905
Magnesium	Mg	12	24.312	Zinc	Zn	30	65.37
Manganese	Mn	25	54.9380	Zirconium	Zr	40	91.22
Mendelevium	Md	101	[256]*				

†1 AMU = 1.6605×10^{-27} kg = $(1/12) \times$ (mass of C^{12} atom).

*The numbers in brackets are mass numbers of the most stable isotope of those elements. These elements are radioactive.

TABLE VII. STANDARD CIRCUIT SYMBOLS

Ammeter	(A)	Lamp	
Antenna		Motor	(M)
Battery		Resistor	
Capacitor	or	Switch (double-throw)	
Capacitor (variable)	or	Switch (single-throw)	
Cell		Transformer	
Coil		Transistor (PNP)	
Conductor		Triode Vacuum tube	
Generator (A.C.)	(G)	Voltmeter	(V)
Generator (D.C.)	(G)	Wires (connected)	
Ground	or	Wires (crossing)	

TABLE VIII. ELECTROMAGNETIC SPECTRUM

Source of Energy	Nuclear reactions	Electron transitions in inner orbitals	Electron transitions in outer orbitals	Molecular vibrations	Molecular rotations	Radio transmitter and electrical oscillators			
Electromagnetic Spectrum	Gamma rays	X-rays	Ultraviolet / Visible	Infrared	Radar and microwave	TV and FM	Shortwave	Standard broadcast	Low frequency radio
Wavelength (in meters)	3×10^{-12}	3×10^{-10}	3×10^{-8}	3×10^{-6} 3×10^{-4}	3×10^{-2}	3×10^{0}	3×10^{2}	3×10^{4}	3×10^{6}
Frequency (in Hz)	1×10^{20}	1×10^{18}	1×10^{16}	1×10^{14} 1×10^{12}	1×10^{10}	1×10^{8}	1×10^{6}	1×10^{4}	1×10^{2}
Detectors	Geiger-Müller tubes and scintillation detectors		Photographic film / Phototubes / Eye	Thermocouples and thermistors	Radio receivers				

TABLE IX.   ASTRONOMICAL DATA

1 Light year (L.Y.)	$= 9.46 \times 10^{15}$ m
1 Parsec (pc)	$= 3.08 \times 10^{16}$ m
1 Astronomical unit (A.U.)	$= 1.50 \times 10^{11}$ m
(Earth-Sun distance)	
Radius of Sun	$= 6.96 \times 10^{8}$ m
Earth-moon distance	$= 3.84 \times 10^{8}$ m
Radius of Earth	$= 6.38 \times 10^{6}$ m
Radius of moon	$= 1.74 \times 10^{6}$ m
Mass of Sun	$= 1.99 \times 10^{30}$ kg
Mass of Earth	$= 5.98 \times 10^{24}$ kg
Mass of moon	$= 7.35 \times 10^{22}$ kg
Average orbital speed of Earth	$= 2.98 \times 10^{4}$ m/s

TABLE X.   PHYSICAL CONSTANTS

Velocity of light in vacuum	$c = 2.998 \times 10^{8}$ m/s
Charge of electron	$e = 4.80 \times 10^{-10}$ statC
	$= 1.60 \times 10^{-19}$ C
	$e^{2} = 1.44 \times 10^{-15}$ MeV-m
Planck's constant	$h = 6.63 \times 10^{-34}$ J-s
	$= 4.14 \times 10^{-15}$ eV-s
	$= 4.14 \times 10^{-21}$ MeV-s
	$\hbar = 6.58 \times 10^{-16}$ eV-s
	$hc = 1.24 \times 10^{-6}$ eV-m
Boltzmann's constant	$k = 1.38 \times 10^{-23}$ J/°K
	$= 0.862 \times 10^{-4}$ eV/°K
Avogadro's number	$N_{o} = 6.022 \times 10^{23}$ mole^{-1}
Electron mass	$m_{e} = 9.11 \times 10^{-31}$ kg
	$m_{e}c^{2} = 0.511$ MeV
Proton mass	$m_{p} = 1.6726 \times 10^{-27}$ kg
	$= 1836.11 \ m_{e}$
	$m_{p}c^{2} = 938.259$ MeV
Neutron mass	$m_{n} = 1.6749 \times 10^{-27}$ kg
	$m_{n}c^{2} = 939.553$ MeV
Atomic mass unit	1 AMU $= 1.6605 \times 10^{-27}$ kg
	$(1 \ \text{AMU}) \times c^{2} = 931.481$ MeV
Gravitational constant	$G = 6.673 \times 10^{-11}$ N-m²/kg
Gas constant	$R = 1.986$ cal/mole-°K
	$= 8.314$ J/mole-°K

APPENDIX B.   ANSWERS TO EXERCISES

1.   6 cm

2.   2500

3.   −0.0175

4. $\vec{R}_F'$ is a force of 4.47 N directed at an angle of 63°4 with respect to the positive X-axis.

5.   0.0268 m/s

6. 0.987×10^{-7} cm

7.   Zero

8.   3.275 g

9.   30.99 m/s

10. 53°1

11. 45°

12. 0.707×10^7 g-cm/s

13. 45°

14. $\vec{R}_F$ is a force of 2.83 N directed at an angle of 45° with respect to the positive X-axis.

15. 60 dynes

16. 4.80×10^{-10} statC

17. 0.063

18. 500 dynes

19. $y_{C.M.} = 4$ cm.

20. 4

21. 81

22. 20 N

23. 45°

24. 4.68 g

25. A velocity of 60 m/s directed 45° south of due west.

26. Down

27. 10.0 dynes

28. A velocity of 316 mi/hr directed 18°4 east of due north.

29. 10^{11} kg-m²/s

30. 0.0175

31. 3.78 m

32. 3.274 m, 3.278 m

33. 1.23×10^5

34. 1.4 Cal

35. 0.8 J

36. 2

37. 0.289

38. 23.4

39. Zero (the null vector)

40. $y = \dfrac{4}{7} x + \dfrac{38}{7}$

41. 3.617×10^5 cm

42. 4 cm

43. $\dfrac{3}{4}$ (0.750)

44. 16°

45. 0.500

46. $\theta = 123°7$ measured from the positive X-axis.

47. $\dfrac{\pi}{4}$

48. 50 dynes

49. All except the second; the last one.

50. 2.2×10^3 lb

51. π

52. 10.0 dynes

53. 7.3×10^{-5} rad/s

54. 1

55. −6

56. 784 J

57. N-m²/kg²; dyne-cm²/g²

58. $\sqrt{61} = 7.81$ units

59. $x = 6, y = -2$

60. 35°1

61. Zero

62. 1.428

63. 135°

64. Triangle (or two straight lines)

65. −0.966

66. 4.9 m

67. 3.10×10^{41}

68. 4.06×10^{-3}

69. 4.83×10^3 g

70. 4.7 ± 1.8 g $(5 \pm 2$ g)

71. $\frac{4}{3}\pi = 4.2$ kg

72. 640

73. Positive

74. 1.3 kg

75. 10^7

76. $\dfrac{n(n-1)(n-2)(n-3)}{1\cdot 2\cdot 3\cdot 4}\,b^4$

77. $\pm 2\% = \pm 0.16$ cm²

78. 0.800

79. 1.50×10^{-2}

80. 125 cm³

81. $x = (a/c) - b$

82. 40 cm/s

83. 114 dynes

84. $-0.5\,\vec{F}$ is a force of 7 N directed vertically upward.

85. 0.903

86.

Time t (seconds)	Displacement x (feet)
0	0
0.5	27.5
1.0	55.0
1.5	82.5
2.0	110.0
2.5	137.5
3.0	165.0
3.5	192.5
4.0	220.0
4.5	247.5
5.0	275.0

87. 1,000,000,000

88. 2.7×10^{34} kg-m²/s

89. 5000 cm/s

90. 18°

91. 10^{-6}

92. 1.4 g/cm³

93. 1

94. 64

95. 15°

96. 1.414×10^5 cm

97. 20

98. 1.728

99. $-1 + \dfrac{1}{2} - \dfrac{1}{3} + \dfrac{1}{4} - \dfrac{1}{5} + \cdots$

100. −5.145

101. 2×10^6

102. $10^8 = 100,000,000$

103. $x = \dfrac{13}{4}$

104. 10^{-6}

105. $a = -400 \cos 20t$ cm/s²

106. $x = v_0 t$ where $v_0 = 100$ km/s

107. 11

108. 0.600

109. 3

110. 57.6

111. −3.078

112. 0.0175

113. 8.75

114. 8.3 light min

115. 10^{12}

116. The displacement x (in feet) increases with time t (in seconds) at a constant rate of 55 ft/s.

117. Zero

118. $>, \cong$

119. 4.518 s

120. 0.0175

121. 1.69

122. Zero

123. Zero

124. $\pm 6\% = 0.06$ cm³

125. 2.38×10^4 ft $= 4.50$ mi

126. $x = 8$

127. $26°6$

128. $\dfrac{4}{7}$

129. 8.49 cm

130. $100\pi = 314.16$ kg

131. 0.15

132. 0.052

133. $53°1$

134. $-15\,\vec{F}$ is a force of 210 N directed vertically upward.

135. $(1.77 \pm 0.05) \times 10^3$ J

136. 0.505

137. Square

138. 8.58×10^{-2}

139. $\dfrac{16}{9}$

140. A displacement of 26.5 mi directed $40°9$ north of due east.

141. 1.5×10^8 ergs

142. 11

143. $<$

144. $\pm 2.4\% = \pm 2.5$ cm²

145. 6.45 ± 0.02 kg

146. 10^3

147. 0.951×10^8 cm/s

148. -1

149. 9 m

150. 107.1 ± 0.8 cm

151. 1

152. 5.385 cm

153. $1 - 0.222 \times 10^{-2} = 0.9978$ m

154. Up

155. 8

156. 0.301

157. 0.043 s

158. 29.4 m/s

159. 7 s

160. $\sqrt{65} = 8.06$ units

161. 86.6 dynes

162. 50.4

163. 2.25×10^4

164. $\dfrac{\pi}{10}\,s\ (0.314\,s)$

165. $-4\,\vec{x}$ is a displacement of 100 mi directed 40° south of due east.

166. 1

167. Since 307 lies far outside the range $421 \pm \sqrt{421} = 421 \pm 20$, the conditions for the two counts were probably different; perhaps the second day was a holiday.

168. 9°

169. 75°

170. Positive

171. $\sqrt{\dfrac{5}{9}}\ (0.745)$

172. $13\,\vec{F}$ is a force of 182 N directed vertically downward.

173. 320 N-m

174. 1 cm

175. 14

176. $5000\,\vec{F}$ is a force of 70,000 N directed vertically downward.

177. 1.7×10^3 cm²

178. A velocity of 420 m/s directed 45° north of due east.

179. $x = -\dfrac{19}{6}, \dfrac{1}{2}$

180. $58°2$

181. 50 s

182. 10^{-9}

183. 4.07×10^{18} cm

184. 1.698

185. $16\pi = 50.3$ cm^2

186. $x = 4, y = 9$

187. 2.21×10^{25} cm^3

188. $270°$

189. 6 cm

190. Negative

191. 11.3 kg

192. 8.660 in

193. 1.414 ft

194. $x = -2.4, 2$

195. 10^{-7}

196. 0.664×10^{-15}

197. 0.075

198. 3.29×10^4 dyne-cm

199. ~ 360

200. 5 cm^2

201. $\dfrac{5}{3}$

202. 6.24×10^4 m

203. 1.333

204. 3.00

205. 0

206. 27.2

207. -1.333

208. $0.005 = 0.5\%$

209. 6.87×10^8 m

210. Positive

211. 44.27 m/s

212. A velocity of 120 m/s directed 45° south of due west.

213. $0.0008 = 0.08\%$

214. 8 cm^3

215. $1 + \dfrac{1}{3}x$

216. -0.454

217. $y = x + 2$

218. 10^6

219. $P = \dfrac{W}{t} = \dfrac{Fs}{t} = Fv; \ v = 20$ m/s; $P = 400$ W

220. $0.001 = 0.1\%$

221. $0°$

222. -500 dynes

223. 10^{-26} gm-cm^2/s

224. 1.58

225. $y = \dfrac{5}{3}x$

226. It is possible to represent an *event* in a conventional plot only if one of the space coordinates is suppressed (leaving, for example, x, y, and t).

227. $\sim 10^{57}$

228. 2.09×10^3 mi

229. 2.59×10^{10} cm^2

230. 10^7 g-cm/s

231. $\theta = 45°$ measured from the positive X-axis.

232. 4×10^3

233. 36.67 m/s

234. 42.4 cm^2

235. $x = 6$

236. 1.204

237. 0.954

238. 10^8 cm/s

239. 2.75×10^2

240. 8.7 ± 1.2 m

241. 20.70 mi

242. 1.36

243. 45 ± 1 mi/hr

244. $3.156 \times 10^7 \cong \pi \times 10^7$

245. 2.66×10^{-26} kg

246. 48°3

247. $x = \dfrac{1}{8}(a + b + 7)$

248. 1.6×10^{-13} J

249. 30 cm/s

250. 1.67×10^{-24} g

251. -0.9998

252. Zero

253. $x = 4, 5$

254. 3.79×10^3 cm³

255. 3.67

256. $v = -20 \sin 20t$ cm/s

257. 18.0 N

258. 0.56 rad/s

259. $y = 3$

260. $\dfrac{5}{9}$ (0.555)

261. 2.07×10^{13}

262. 2

263. Up

264. 10.42 mi

265. 50 dynes

266. 7.68 cm

267. $1 + x$

268. $t = -5$ s, 1.25 s

269. 41

270. 63.2 g

271. -0.707×10^7 g-cm/s

272. Down

273. 5.48×10^2

274. 10^5

275. 3

276. $1 + \dfrac{3}{2}x + \dfrac{3}{8}x^2 - \dfrac{1}{16}x^3 + \cdots$

277. 6.6

278. 147 rad/s

279. 2.45

280. 2

281. 25 dynes

282. 10^3

283. $\dfrac{\pi}{4}$ radians

284. 30°

285. $3\vec{x}$ is a displacement of 75 mi directed 40° north of due west.

286. 8.64 ± 0.06 g

287. 45°

288. A velocity of 240 m/s directed 45° north of due east.

289. The displacement x (in kilometers) increases with time t (in seconds) at a constant rate of 100 km/s.

290. $x = \cos 20t$ cm

291. 1.3 pc

292. 3.5×10^4 dyne-cm

293. 10^{-5}

294. 90 rad/s

295. $x = \dfrac{1}{2}$, $y = 2$

296. 0.105

297. 0.398

298. 0.29 mi/s

299. $y = -x + 49$

300. 1.76

301. $-1.5\vec{x}$ is a displacement of 37.5 mi directed 40° south of due east.

302. $\dfrac{1}{\sqrt{2}}$ (0.707)

303. $6\,\vec{x}$ is a displacement of 150 mi directed 40° north of due west.

304. 1.87×10^7 m/s

305. $1 - \dfrac{3}{2}\,x + \dfrac{15}{8}\,x^2 - \dfrac{35}{16}\,x^3 + \cdots$

306. 1.04

307. 6.75 dynes

308. $\cong$

309. 4.328

310. 10^7

311. 0.334

312. 0.62

313. 36.67 m/s

314. $y = \dfrac{16}{9}\,x - \dfrac{47}{9}$

315. 17.1 ± 0.3 m

316. 0.061

317. -0.364

318. 8 m

319. 5 m

320. 171°9

321. 5 mi

322. 12

323. 2.5×10^{-6}

324. 0.707

325. 96.6 dynes

326. -1

327. $\propto$

328. $876 - 936$

329. 0.800

330. $0.01 \text{ rad} = 0°57$

331. 0.707

332. 5 mi

333. 55.9 dynes

334. 15 cm/s

335. $\sqrt{48} = 6.93$ m

336. $\pm 3\% = \pm 27$ m²

337. 2.16

338. 0.954

339. 50 cm/s

340. No

341. 36°9

342. $16\pi = 50.3$ cm³

343. 4.3274 kg

344. $(-2 + \sqrt{14})$ s $\cong 1.742$ s

345. Up

346. 25 dynes

347. 2×10^{-3} C

348. 41.9 cm³

349. Rectangle

350. 50 dynes

351. $\sim 10^{18}$ s

352. 0.309×10^8 cm/s

353. 9.94

354. Straight line

355. 1.28

356. 1.009

357. $\gtrsim$, $\approx$

358. Positive

359. $x = -1, 3$

360. $R = 37$ mi; $\vec{R}$ is directed 22°5 north of due east.

361. 6.75×10^{-5} N

INDEX

229